GUT REACTIONS

THE SCIENCE OF WEIGHT GAIN AND LOSS

SIMON QUELLEN FIELD

CHICAGO REVIEW PRESS

Published by Chicago Review Press Incorporated
814 North Franklin Street
Chicago, Illinois 60610
ISBN 978-1-64160-000-2

Library of Congress Cataloging-in-Publication Data
Names: Field, Simon (Simon Quellen), author.
Title: Gut reactions : the science of weight gain and loss / Simon Quellen
    Field.
Description: Chicago, Illinois : Chicago Review Press Incorporated, [2019] |
    Includes bibliographical references and index.
Identifiers: LCCN 2018039859 (print) | LCCN 2018050143 (ebook) | ISBN
    9781641600019 (PDF edition) | ISBN 9781641600033 (EPUB edition) | ISBN
    9781641600026 (Kindle edition) | ISBN 9781641600002 | ISBN
    9781641600002 (trade paperback edition)
Subjects: LCSH: Weight gain. | Weight loss.
Classification: LCC RM222 (ebook) | LCC RM222 .F49 2019 (print) | DDC
    613.2/5—dc23
LC record available at https://lccn.loc.gov/2018039859

Cover design: John Yates at Stealworks
Cover images: (donut) Ragnar Schmuck/Getty Images; (beaker) urfinguss/iStock
Interior design: Jonathan Hahn

Printed in the United States of America
5 4 3 2 1

CONTENTS

INTRODUCTION

Walk past the diet section in a bookstore and you will find hundreds of books that each claim to have the solution to weight problems. Whether it's a low-fat diet, a low-carbohydrate diet, a Paleolithic diet, a gluten-free diet, or any number of other magical things to try, they all claim that your problem is simple and has a simple solution.

The popularity of these diets is linked to the observation that they all seem to work. Why would these different approaches to weight control all have success? One reason is that they all limit your choices. When you are not able to eat everything in sight, you tend to eat less. Another reason is that each diet targets a different set of people and works for that set, if not for others.

Of course, they all work until they don't. Limiting your diet in this way is difficult to do for long periods of time.

There aren't many things that we expect to work for all people all the time. People are different from one another. There are genetic differences, cultural differences, behavioral differences, psychological differences, and social differences, and they all play a part in the weight maintenance equation. Add to these differences age, gender, gut microbiology, and environment and it becomes clear that weight control is a personal thing, and the solutions might have to change not only with each individual, but also with time and setting.

Obesity is not one disease. There are many subtypes of obesity. Some people will have disorders in how their bodies manage energy balance, while others may have differences in how they respond to reward, making it hard for them to resist highly palatable foods. There are many ways to affect our body composition toward more fat, and that is what this book is about.

HOMEOSTASIS

Homeostasis is the word used for the body's automatic system for weight control. Homeostasis means "staying the same" (from the Greek words for "similar" and "standing still"). It is very important for the survival of any organism. Without it, the animal starves to death, or feeds to death. Biological systems that are this important have many components, with several checks and balances, and interactions with other systems. Things get complicated.

There are two kinds of weight problems that are related to homeostasis.

Some people slowly gain weight all their life. A weight gain of a single pound a year adds up over time, and by age 40 or 50, those 20 to 30 pounds of extra weight are noticeable in the mirror, on the scale, and in the doctor's office.

Other people find that they reach a particular weight and stay there, in homeostasis, but at a weight that is unhealthy or otherwise undesired.

In the first case, homeostasis isn't working properly. In the second case, the controls are set higher than we'd like, and we want to change them.

We keep many of our body systems in homeostasis. We regulate our temperature—we can die if we get too hot or too cold. We regulate our blood pressure and our blood sugar. Again, we can die if either gets too high or too low. Problems with blood sugar homeostasis cause diabetes. It turns out that all these other control systems interact with the control systems for body fat. Disease and fat can be correlated. Obesity is a risk factor for diabetes, and raising the body temperature during a fever burns fat.

HISTORY

As I write this, over 68 percent of adult Americans are overweight, and over 35 percent are obese. We know this isn't *just* a little bit of weight gain over many years because 32 percent of children and teenagers are over-weight, and 17 percent are obese. They haven't had time to accumulate their fat slowly. But since obesity rates do go up with age, some part of the problem is small increases over time.

If these numbers had always been the same, we wouldn't worry. We don't worry about obese elephant seals or hippos. Their fat percentages are normal for their species and are essential for their survival. But humans have been gaining weight since the 1960s. While the percentage of people who are overweight but not obese has remained fairly stable, the percent-age of obese people has risen from about 13 percent in the 1960s to over 35 percent today (Ogden and Carroll 2010).

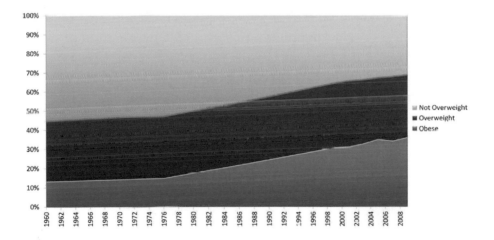

Something happened between 1980 and now. The sharp corners in the graph above are an artifact caused by the fact that the data comes from surveys done at particular times. Nonetheless, the overall trend is real. What changed in those years that caused the rise? What has caused the rates to stabilize since 2010? We will look into those questions in the fol-lowing pages.

The fact that obesity rates changed in the late 1970s, and then stabilized in the 2000s, is actually good news. It means that obesity rates can change. What was done can be undone. In fact, the recent stabilization in growth is likely due to new scientific knowledge about fat homeostasis, and as we make that information more widely available, there is hope that obesity rates may fall as quickly as they rose.

In preindustrial America (before 1900), the obesity rate hovered around 4 percent, compared to 35 percent today.

Some of the changes in the American diet since that time are related to the diversity of the foods available and the ease of preparing food. Before supermarkets, kitchen stoves, refrigerators, restaurants, fast food, and frozen dinners, the variety of foods available was limited. Limiting choice in food is a form of dieting. Food becomes less interesting, and we eat when we are hungry, rather than when we are tempted.

Another change in diet was the rise in consumption of polyunsaturated seed oils, rich in omega-6 fatty acids. We are eating three times the amount of polyunsaturated oil as we were before 1900. Also striking is our doubling of sugar consumption in that period. Consumption of sweetened beverages is five times what it was in 1900.

Our food technology has also allowed manufacturers to improve the palatability of our food, at the same time they improve the appearance, shelf life, and manufacturing efficiencies. Since 1970, Americans have increased the amount eaten per day by 425 calories, a full 20 percent, which may be due to a combination of a wide variety of foods available, the high palatability of those foods, the ease of eating them at a restaurant or out of the microwave, and other factors we have not (yet) discussed.

If we are eating 20 percent more food than we were 45 years ago, is that where all the fat is coming from? Not necessarily. The natural homeostasis systems in our body would normally do something with that food other than store it as fat. We would increase our metabolic rate and our level of activity. The fat cells would produce signals that tell us we are full. Our stomach would stop producing hunger signals. Whether this is enough to overcome the temptation of eating all that delicious food that is sitting right there in front of us is something we will explore further later.

If we eat 20 percent more food than we need, and it goes into making new fat cells, those cells will need food to survive. If we gain weight, we will need more food just to move all that weight around all day. We can't just say that adding 425 calories per day adds some fixed amount of fat every day. We should expect that the amount of added fat would decrease over time. The same math works in reverse. Dieting should eventually lead to a plateau in weight, as the lost calories will eventually lead to a lower metabolic rate and less activity, accompanied by an increase in hunger signals that causes you to cheat on the diet.

The amount of exercise people got in the early 1900s (when the obesity rate was 4 percent) is also likely more than what the average American is getting today. Transportation back then involved more walking. Work involved more moving around than the typical desk job does today. Entertainment did not involve sitting in front of a screen at home. While these differences in the way society works and plays are not likely to change, individuals can choose to get more exercise, and we have tools today that can monitor the number of steps we take and remind us to get up and walk around. I would not, however, expect that exercise is the answer to weight loss. We burn nearly all our calories just staying warm, pumping blood, and thinking. If you were to go running for an hour a day, the extra calories you burn would be at most about 20 percent of your daily calories. Then homeostasis would kick in, and you would relax for the rest of the day to make those calories up.

We are also getting less sleep than people used to get. Sleep is important for regulating metabolism, and lack of sleep makes us hungry in two different ways, which makes us eat more. (More on this later.) Sleep can be disturbed as we fool our diurnal clocks by using artificial lighting and watching television, computers, or cellphone screens late into the night. In particular, getting blue light after sunset can cause sleep abnormalities, and there is a lot of blue light in our various electronic media—you can't make white pixels on a screen without blue.

Where you live can make a difference in your diet and the amount of exercise you get. At the top of the next page is a "heat map" of US counties, colored by the percentage of obese people in the population.

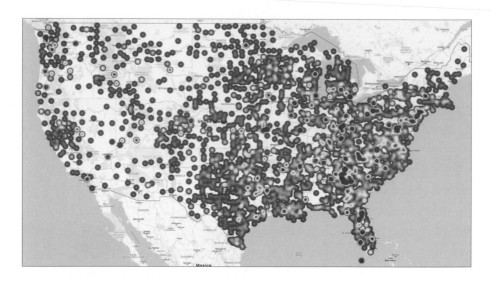

There are clearly "hot spots" in the distribution. This is unlikely to be due solely to genetics, as people in the United States have a high degree of mobility and a diverse population. Likewise, cuisine does not vary enough to account for the distribution. There is a very good national food distribution system, and supermarkets across the country carry much the same kind of food.

What does seem to correlate to a good extent is poverty. The United States subsidizes crops like corn, wheat, and sugar, and these high-carbohydrate

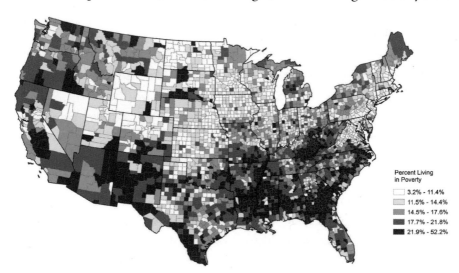

foods are easy to process into cheap addictive products with long shelf lives, easy transportation and storage, and lots of marketing.

In contrast, fresh fruits and vegetables are not easy to find in poor neighborhoods, and animal protein sources are expensive. Education about diet is also an issue in areas with low education levels or where language barriers exist. Compare the locations of obese counties in the heat map to the high-poverty counties in the map on the bottom of the previous page.

We see the same trends with other addictive substances, such as alcohol, tobacco, and recreational drugs.

EATING FOR REWARD

Two changes in lifestyle have also contributed to the obesity problem. We have become very good at designing foods that are highly palatable. We joke about these foods being addictive, but it turns out not to be a joke. These foods stimulate the same reward circuits in the brain that other addictive substances do. We are eating for pleasure, long after we have satisfied our hunger. The satiety signals our homeostasis system is sending us are overridden by the reward. After a meal, we might turn down a second helping of broccoli because we are full. But we seem to have room for a dessert that has even more calories. This is partly because we are addicted to the reward, but also because sugar and wheat flour don't produce the satiety signals that other foods do, so they don't tell us to stop eating.

The second change is the easy availability of these highly palatable foods. We don't have to mix and bake our own cookies. They are in that big package in the cupboard. The ice cream is in the freezer, waiting for us. No blending of ingredients and hand cranking the ice cream maker. Wherever we go, there are vending machines full of candy bars. At every checkout counter, even at the hardware store, there is a rack of cookies and candy bars, and maybe even a freezer full of ice cream bars. The temptation to feed our addiction is ever present.

Someone addicted to alcohol and nicotine can avoid smoky bars, but everyone needs to shop for things, and all those delicious calories are there at the checkout counter at the bookstore, the movie theater, the gas station, the toy store, and even the car wash. Many places have these foods available for free. Banks often have cookies or candy on a table by the door and at the tellers' windows. Car repair shops have an array of pastries next to the free coffee. Gyms and health food stores sell energy bars and protein bars that are basically candy by another name.

Highly palatable foods act on the brain's reward center. This bypasses the homeostasis system in the brain. The homeostasis system will tell us we are no longer hungry. The reward system will tell us that hunger does not matter because we are eating for pleasure, not to satisfy hunger. It is the reward system that makes people addicted to nicotine, opiates, cocaine, and amphetamines.

Highly palatable foods have been designed to be addictive. They contain sugar, salt, and fat, in an irresistible combination. They require almost no chewing. They melt in the mouth. They have a variety of flavors and aromas, usually in the same bite. They are highly advertised and are easily available, and ubiquitous, because they sell so well. Just like any other addictive substance.

The brain becomes addicted by training. We are not born craving nicotine or chocolate chip cookies. We gradually program our brains to need these things, instead of just wanting them. We lose control. And as with all addictive substances, we can regain control by consciously changing our relationship with the substance. Far fewer people smoke cigarettes in the United States than in the 1950s. What used to be cool and attractive is now seen as a disgusting habit by many. If we decide to eat only when we are hungry, to think about how we will feel after overeating or what the scale will read the next morning, we can gradually reprogram the brain away from addiction and regain control over our eating habits.

Not everyone who is overweight is an addict, has a damaged homeostasis system, sleep disorders, bad genes, bad eating habits, neuroses, gut microbe problems (dysbiosis), unhealthy environment, or any particular one of the many reasons people are fat. But most of us, even those who are not fat, have at least one and probably several of those issues. We are all different, and to take control of our bodies, it helps to understand what our reasons and triggers are and how to overcome them. What works for one person might work for many or most but might not work for all people.

WHY ARE WE FAT?

As a nation and, increasingly, as a planet, we're fat. Why is that?

It could be the types of food we eat. Early on, we thought the problem was the fat in our diet. We made a big deal about fat and created low-fat this and low-fat that, and we got fatter. Then we thought that maybe it was carbohydrates. Various diet plans that reduced carbohydrate consumption seemed to work for those who tried them and could stick with them. But we still got fatter as a nation. There are people now claiming that the problem is with the last of the three macronutrients: protein. We eat too much meat, they say. Maybe. We certainly eat much more protein than we need. We'll look into all these ideas in a bit.

Are we getting less exercise than we used to? Probably. That seems likely to be part of the problem, at least for some people. We work at desks now instead of plowing fields and roping cattle. We drive instead of walking or riding horses. We have laborsaving devices to do most of the work we used to do. But of the 2,000+ calories we eat each day, exercise doesn't actually burn that many. Most of the calories go into keeping us warm, keeping blood flowing, keeping us breathing, digesting, and thinking—things we do even when we aren't exercising.

We get fatter as we get older, and as a nation, we are getting older.

Maybe there's something in the environment that causes us to overeat. Advertising. Cheap and abundant food. A wide variety of food. We eat out at restaurants more than we used to. There are bigger portions at restaurants. Food has gotten more addictive as the makers of processed foods have learned the science of creating foods we can't pass up and can't stop eating.

Maybe it's in our genes. Four-fifths of body mass index (BMI) is heritable. If your parents were fat, you are likely to be fat. But is that because you are in the same environment as your parents, or because they taught you bad eating habits?

Personality plays a part. Conscientious people can stick to diets better, and more neurotic people cannot. Are we less conscientious and more neurotic than we used to be?

Perhaps we traded one vice for another. People are smoking a lot less these days, and people may have moved from an addiction to nicotine to an addiction to sugar, fat, and salt.

We are finding out that our symbiotic gut bacteria have a lot to say about how much we eat and how well we digest it. Changing the gut bacteria in mice can make lean mice fat and fat mice lean, and there are studies in humans showing similar effects. Have we changed the ecology in our lower intestines? There is a lot of research showing that the typical American diet does just that, and in a way, that makes us fatter. But is that everyone's problem? Probably not. For any particular person, there may be several different reasons why keeping trim is a challenge, and those reasons will be different for each person.

THE *TYPES* OF FOOD WE EAT CAN MAKE US FAT

Both low-fat diets and low-carbohydrate diets help people lose weight (Hu et al. 2012). Moreover, the amount of weight loss and the amount of waist circumference loss is generally about the same. If one or the other macronutrient—fat or carbohydrate—were the cause of obesity, we would not expect equal results when reducing either one while keeping total calories constant.

There can be several reasons why both diets seem to work. By limiting the types of food the individual is allowed to eat, we reduce the variety. Both fat and carbohydrates (especially sugar) increase the palatability of food, improving its taste and making it more rewarding. Reducing either one makes food less appealing. Even when allowed to otherwise eat freely, people on low-fat diets did not compensate by adding calories (Kendall et al. 1991), and they lost weight.

Simply thinking about tasty food causes the body to release insulin in anticipation of the meal. Insulin reduces blood sugar levels, making us hungry. What does the body do with the sugar from the blood? It does not throw it away. It does not burn it. It stores it as fat. If the food is less appealing, it might not be as fattening, even if it contained the same number of calories as a tastier meal.

People vary in how they respond to insulin (insulin sensitivity) and how much insulin they produce after being given the same dose of glucose (insulin concentration). In those with high insulin concentration (an indicator of low insulin sensitivity), a low glycemic load diet caused faster weight loss than a low-fat diet (Ebbeling et al. 2007). Reducing the sugar load reduces the insulin. So while in general we can say a low-carbohydrate diet is as good as a low-fat diet, the results for any particular person can depend on that person's insulin sensitivity, and perhaps other differences as well: genetics, gut microbes, exercise levels, etc.

SUGARS AND OTHER CARBOHYDRATES

SUGAR

Sugar is everywhere in the American diet. A 12-ounce can of sugared soda has 39 grams of sugar. To get that same amount of sugar, you'd have to eat 3½ cups of unsweetened applesauce. This may be why most applesauce sold in the United States has been sweetened to double the amount of sugar (SugarStacker 2016). While we know that there is sugar in soft drinks, sugar is added to almost everything, from ketchup to salad dressing, and it can be hard to avoid, or even to know when it is being consumed.

It is difficult to consume too much sugar from easy-to-find sources in nature. Our ancestors would have found some sugar in fruit, where it is packaged with water bound to fiber, so it is quite filling. And their access to fruit was limited by the seasons; in the winter and spring, they would have found it quite hard to come by. In the supermarket, on the other hand, what passes for fruit outside of the produce section is often some highly processed treat whose first three ingredients are forms of sugar, such as the fruit snacks whose ingredient labels are shown on the next page.

Sugar is an easy way to get calories, especially when it is in liquid form, as it is in soft drinks. No chewing is required, no preparation, no cooking—just the sweet stuff. A teenager who would be hard pressed to

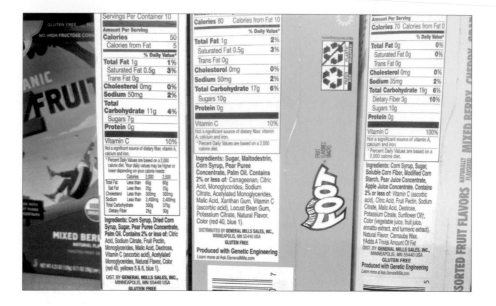

consume 6 or 9 cups of applesauce can easily consume two or three cans of sugared soft drink at a meal or a party. But there is more to the problem of sugar than that. Sugar interferes with the homeostasis systems in the body. It raises blood sugar levels quickly. This is a dangerous condition, and the body responds by sending insulin into the blood. Insulin's job is to tell the fat and muscle cells to remove the sugar from the blood by storing it as fat. After this rush of insulin, the blood is now low in sugar, making you hungry again. Therefore, you eat more of what was in front of you. More sugar.

After we eat most unprocessed foods, glucose is released into our blood slowly, and insulin levels don't spike sharply. There is enough time for the insulin to tell the brain we aren't hungry anymore, and we stop eating. The blood sugar levels don't suddenly plummet; they remain regulated. Insulin decreases the production of the hunger-stimulating hormone ghrelin. Insulin increases production of the satiety hormone leptin, making us feel satisfied (satiated) (Teff et al. 2004).

Fruit juices are just as bad as soda. The fiber has been removed from the fruit, so it is not there to fill the stomach and tell us we've had enough. These juices are little more than sugar water. What tiny amount of nutrition is left in the juice is not significant, and almost anything you substitute for the juice, such as whole fruit, is likely to be more nutritious.

Sugar is an addictive substance (Avena, Rada, and Hoebel 2008). Sugar releases opioids and dopamine into the brain, just like opiates and other drugs such as nicotine and cocaine. The pleasure and reward systems in the brain are stimulated. Excessive sugar intake (sugar bingeing) is one indication of addiction, as are withdrawal, cravings, and "cross-sensitization" (increased sensitivity to stimuli that accompany the addictive drug) (Avena and Hoebel 2003). People who are more prone to addictive behavior—alcoholics, ex-alcoholics, smokers, ex-smokers, etc.—are more prone to addiction to highly palatable foods and exhibit those same four indicators (Davis 2013).

Sugar is hidden in most processed foods. In some, it is listed plainly as the first ingredient, as it is in the package of raisin pastry shown below.

This is clearly a dessert, although it was being served as a breakfast in the hotel I recently visited. *Malted barley flour* is another term for sugar. When barley is allowed to germinate, in a process known as malting, it produces enzymes that convert the starch into glucose, a simple sugar. Other sugars mentioned on the ingredients label are dextrose (another word for glucose), corn syrup, and corn syrup solids.

Another dessert masquerading as a breakfast is sweetened cereal. When sugar is the second ingredient, what you are eating is dessert. But in many cases, sugar is also the third, fourth, and fifth ingredient, under slightly different names. If the peanut butter, dextrose, corn syrup, brown sugar syrup, and molasses were added to the sugar instead of listed separately, we might see sugar properly labeled as the first ingredient.

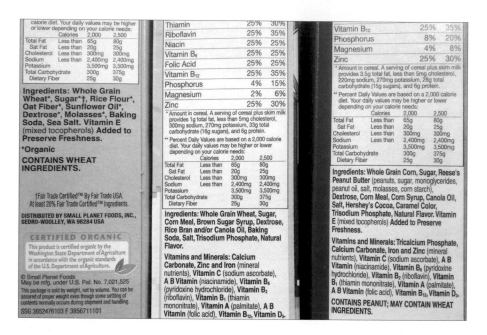

Other names for sugar found in processed foods include invert sugar, invert cane syrup, tapioca syrup, malt flavor, or as many as 100 different names.

A single sugar-sweetened soft drink per day is enough to measurably increase childhood obesity and increase body mass indexes (BMI, a proxy for body fat fraction) in children (Ludwig, Peterson, and Gortmaker 2001).

There is good news for anyone considering giving up sweets, including artificial sweeteners. It turns out that after a couple months of little or no sugar in the diet, the brain and tongue reprogram themselves to detect much lower amounts of sweet. Things that never tasted sweet before (because they were being drowned out by the huge sugar load in the modern diet) now taste sweet (Wise et al. 2015).

In nature, sugar is usually found in the form of sucrose, a molecule composed of two simple sugars, glucose and fructose.

The gray atoms are carbon, the red atoms are oxygen, and the white are hydrogen. We call them carbohydrates because they are made of carbon and water (you can count the hydrogens to see there are two of them for every oxygen).

Your body has to use a tiny bit of energy to break the sucrose molecule apart into the two simple sugars. The glucose can be used immediately by the muscles, brain, and other organs. The fructose is sent to the liver for processing.

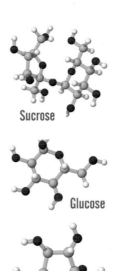

Sucrose

Glucose

Fructose

Fructose

Fructose does not raise insulin levels. Thus, the appetite-stimulating hormone ghrelin is still produced, and we still feel hungry. The satiety hormone leptin is not produced, so we don't feel full. The glucose receptors in the brain's hypothalamus that tell us to stop feeding are not activated. The tasty food is still in front of us, and our homeostasis mechanisms aren't kicking in to stop us from overeating.

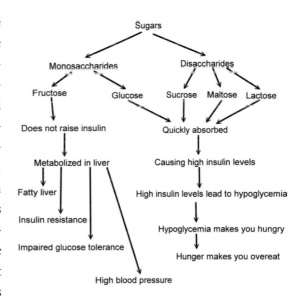

Eating a high-fructose meal leads to high levels of fat (triglycerides), even when the same total number of calories is controlled for (Teff et al. 2004). About 30 percent of the fructose is turned into fat in the liver. Much of this extra fat is stored in the liver, where it was produced, not in fat cells elsewhere in the body. Excess liver fat causes the same kind of liver disease as excess alcohol consumption. It is called "nonalcoholic fatty liver disease." Fat in and around the liver (known as visceral fat, abdominal fat, or "beer belly") is the dangerous fat associated with heart disease, stroke, and diabetes.

Not all the fat made from the fructose is stored, however. Some of it leaks out into the blood as free fatty acids. Those cause the muscle cells to become less sensitive to insulin. They no longer hear the signal to burn glucose, so the glucose must be removed from the blood by the fat cells, which store it as fat.

In the liver, fructose metabolism creates the inflammatory molecule JNK1, which has the effect of interfering with the insulin receptors in the liver cells. The liver becomes insensitive to insulin, so it does not hear the insulin signal telling it to stop converting stored glycogen into blood sugar. Blood sugar rises, prompting more insulin production. The muscle and liver aren't listening to that signal, so the fat cells have to remove the sugar and store it as fat.

While this book is about obesity, fructose does more damage than simply adding fat. It causes heart disease by making the liver produce too much uric acid (Johnson et al. 2007). Uric acid raises blood pressure by reducing nitric acid production. (Nitric acid is the molecule that makes blood vessels dilate, lowering blood pressure—it is what nitroglycerin pills release into the blood to relieve angina.) Uric acid is what causes the disease gout.

The high insulin levels raise blood pressure. Combined with the uric acid that prevents the blood vessels from getting the nitric oxide signal to dilate, we now have two signals leading to hypertension, or high blood pressure. The high insulin levels also interfere with the brain's leptin receptors. Leptin is the hormone that tells the brain to stop eating. Leptin also reduces the reward signals in the brain that we get from eating, so eating more isn't as much fun. Without the leptin signal, we still get the reward from the fructose, and we keep eating. Eating fructose not only makes us fat, it makes us hungry. We eat more and get fatter still.

Fructose is sweeter than sucrose and much sweeter than glucose. That is why manufacturers of processed food are adding more fructose to their products. High fructose corn syrup is more than half fructose, and the two simple sugars (glucose and fructose) are already broken apart, unlike in natural sugar (sucrose).

Fructose affects the brain differently than glucose does. Fructose increases feeding, while glucose decreases feeding. Instead of relieving hunger, fructose makes us eat more (K. Page et al. 2013). Besides the brain, fructose has effects on visceral fat, blood fats (triglycerides linked to heart disease and stroke), and decreased insulin sensitivity (seen in pre-diabetes and type 2 diabetes) (Stanhope et al. 2009).

Unlike other foods, fructose does not suppress the hormone ghrelin (produced by the stomach to make you feel hungry). It has all the calories of glucose, but it does not relieve hunger.

Like alcohol, the body cannot store fructose, so the liver must deal with it immediately. This means that any remaining carbohydrates, such as glucose, go straight into storage in fat cells. Since table sugar (sucrose) is half glucose, half of the calories go into fat storage instead of being burned in the muscles.

Fructose causes oxidative damage by increasing free oxygen radicals that damage tissues. Free oxygen radical control is why doctors recommend eating antioxidant foods. Fructose causes inflammation that leads to atherosclerosis and diabetes, among other ailments (Saygin et al. 2016).

Fructose is seven times more likely than glucose to produce advanced glycation end-products (AGEs). AGEs are made when sugars like fructose bind to proteins and fats. Our bodies have receptors for AGEs, because they are dangerous. We respond by producing chronic inflammation, which causes kidney damage, fibrosis, atherosclerosis, Alzheimer's disease, obesity, and other complications of diabetes. AGEs also cause high blood pressure by blocking the action of nitric oxide, the molecule that expands blood vessels. That they are called AGEs is fitting, since they cause many of the problems associated with aging. Although fructose is the main culprit, even mildly elevated blood glucose levels result in higher AGE levels.

The process for making high fructose corn syrup was invented in 1957, and since then manufacturers of processed food have been increasing the amount of it in use. It is inexpensive, and its sweetness can be easily manipulated by varying the amount of fructose. Some formulas have 42 percent fructose, while others have 55 percent to 90 percent. Americans get 17 to 20 percent of their daily energy intake from fructose. There seems to be a direct relationship between increased fructose consumption and increases in obesity, metabolic syndrome, and diabetes. Other symptoms of fructose consumption are insulin resistance, increased fats in the blood (called dyslipidemia), impaired glucose homeostasis, increased body fat, and high blood pressure. The effects of fructose in the diet are in part due to the changes in gut bacteria that a high fructose diet promotes (Payne, Chassard, and Lacroix 2012).

ARTIFICIAL SWEETENERS

Most people who want to lose weight already know they should cut down on sugar or completely eliminate it. But the attraction of sweets makes that difficult. The pleasure we get from sweet things can be so profound as to cause an actual addiction. So how about those calorie-free miracles of science that are so widely available in every diet soda or low-calorie dessert? Can't we just switch to artificial sweeteners, have our cake, and eat it with every meal?

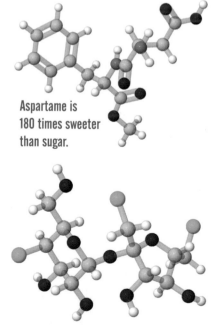

Aspartame is 180 times sweeter than sugar.

Of course, it isn't that easy. Many of the problems with sugar remain with anything that tastes sweet. The sweetness receptors are still sending signals to the brain, telling us to eat more of this delicious stuff, even long after we are no longer hungry. If the sweet food has any calories at all, that leads to overeating. But even if the diet soda is completely calorie free, it still causes problems, simply because it is sweet.

Sucralose is 600 times sweeter than sugar.

Our bodies have sweet sensors in many places, not just on the tongue. We have them in the gut (Steinert et al. 2011), the pancreas (Nakagawa et al. 2009), in other organs, and in the brain. When these sensors detect something they think is sugar, they send their signals to the body to pre-

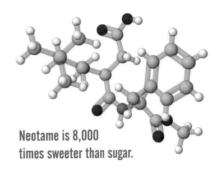

Neotame is 8,000 times sweeter than sugar.

pare for the availability of sugar (Malaisse et al. 1998). The pancreas injects insulin into the blood, which then tells fat cells to remove sugar from the blood and store it as fat. The result is that we have low blood sugar, which makes us hungry. Drinking that diet soda makes us want to eat, even if we were not hungry before we drank it. It makes us convert blood sugar into fat, when without the diet soda, we might have burned that blood sugar to feed our muscles and stay warm (Swithers 2013).

Artificially sweetened beverages are associated with increased body mass index (BMI) and increased body fat percentages, elevated risk of coronary heart disease, elevated risk of hypertension (high blood pressure), as well as double the risk of metabolic syndrome and type 2 diabetes (Swithers 2013).

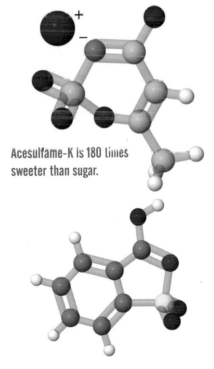

Acesulfame-K is 180 times sweeter than sugar.

Another effect that noncaloric sweeteners have is to reduce the ability of sweet-tasting foods to signal satiety. The body learns that sweet does not mean calories. Therefore, when we *do* get real sugar, we don't respond by eating less (Swithers 2013). When artificially sweetened beverages are consumed along with sugar-sweetened foods, we eat more.

Saccharin is 300 times sweeter than sugar.

Just as with sugar, consuming a lot of artificially sweetened foods and beverages trains the tongue and the brain to ignore the signals of sweetness in foods. Once we give up sugar and artificial sweeteners, after a few weeks, we regain the ability to sense the sweetness in fruits and vegetables, increasing the palatability of healthier foods. Where we might think that we would miss the sweetness of candies and pastries, we regain the delights of natural fruits as compensation.

Artificial sweeteners have yet another downside. They change the composition of our natural gut microbes, reducing our tolerance for glucose (Suez et al. 2014). Because we don't digest artificial sweeteners, they pass directly into the large intestine, where the beneficial gut bacteria live.

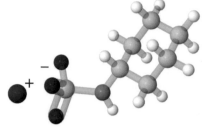

Sodium cyclamate is 30 to 50 times sweeter than sugar.

Any harmful change in gut bacteria is called dysbiosis. One study found that over 40 different types of bacteria changed in abundance with a diet including artificial sweeteners, causing metabolic problems such as those that lead to diabetes and obesity (Suez et al. 2014). The bacteria that were favored by the artificially sweetened diet were the ones that produce the short-chain fatty acids acetate and propionate, which are signal molecules the body uses to control the production of glucose and fats, and which also affect satiety. The types of bacteria favored by the artificial sweeteners are the same ones found in the guts of people with type 2 diabetes and obesity.

STARCH

For hundreds of years before the war on dietary fat began in the 1960s, doctors thought they knew why fat people were fat: too many starchy foods. There weren't that many fat people, but they all seemed to be eating lots of starchy food: bread, puddings, potatoes, cakes, pasta, cereals, and gravy, all with lots of starch. Reduce the amount of starch you eat and you won't be fat—that was the wisdom then.

But things changed. We decided fat was the enemy. We decided that the US Department of Agriculture should be in charge of deciding what we should eat, through a thing called the food pyramid, which placed starchy agricultural products at the base, telling us that a quarter of our plate should be full of starch, which was renamed "complex carbohydrate."

And we got fatter as a nation. We exported our ideas and our agriculture to the rest of the world, and they got fat too.

As more food on the market became processed food, starch took over the supermarket. Most of the center aisles are full of starch. Pasta. Bread. Soups full of potatoes and noodles. Corn chips. French fries. Mashed potatoes. Tater Tots. Hash browns. Cookies. Flour. Rice. Cornmeal. Cornstarch. Tapioca. Pudding. Potato chips. Ramen. Spaghetti. Macaroni. Bakery aisles and whole bakeries inside the store.

Starch lends itself very well to processing. It can be shaped and modified. It can be extruded, formed, and cut by machines that make vast quantities of processed foods in seconds. It can be made crisp or soft. It can thicken liquids to make them more appealing or coat other foods with an attractive brown crust. It gels into a paste, holding water, which adds moist mouthfeel and cheap bulk and weight.

Manufacturers learned to make starch more attractive, appealing, and addictive by further processing. Flour mills now make flour extremely fine, unlike the coarser stone-ground flours of old. The superfine flour allows the starch grains to be rapidly digested into the sugar maltose by enzymes in saliva. The starch has turned into sugar almost before you swallow. The maltose is quickly broken down further by the small intestine, where it splits into two molecules of glucose. The flood of glucose from the quickly digested slice of bread causes a rush of insulin into the blood to handle the overload. The glucose turns into fat, and the blood glucose levels fall, and we are hungry again. Time to buy more manufactured agricultural products. It's almost as if someone planned it.

Types of Starch

There are two types of starch: amylose and amylopectin.

Amylose is a simple chain of glucose molecules all linked together. It usually winds around in a helix shape, like a spring. Sometimes it can be in a single helix shape, as shown below:

At other times, it may be in a double helix shape. In the image below, we have colored one helix red and the other blue:

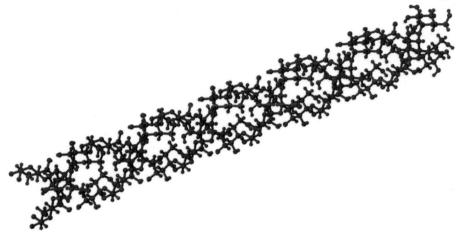

These helixes can stick together with what chemists call hydrogen bonds, and get rather stiff. In a reaction called retrogradation, a solution of amylose in water (a gel) will lose water and start to crystallize. The amylose crystals become much less water soluble, and the enzymes that normally break up starches can no longer reach into the crystal to break apart the glucose molecules. We say that this form of starch is "resistant starch." It acts much like fiber. We don't get calories from it. It goes down into the large intestine, where it feeds the same beneficial gut bacteria that feed on fiber. Those bacteria produce the short-chain fatty acids propionic acid and butyric acid, which are absorbed into the blood and tell the brain we have had enough to eat. Butyric acid in the colon is also believed to be important in preventing colon cancer and other colonic diseases (T. Wang, Bogracheva, and Hedley 1998).

This retrograde crystallized stiff starch is what makes stale bread hard. In fresh bread, the amylose is still in its gel form and can be easily digested by enzymes in saliva. In a hot baked potato, the amylose is also in gel form. But in cold cooked potatoes, such as in a potato salad left in the refrigerator overnight, the amylose has had time to crystallize and become resistant. The portion of the starch that is resistant has effectively zero calories and all the benefits of fiber.

Most sources of starch—wheat, potatoes, rice, etc.—are only about 20 percent amylose. The rest is one of three forms of amylopectin.

Amylopectin is also made up of chains of glucose molecules. However, it has a branched structure that makes it fluffy and easy for the digestive enzymes to access. When cooked, it opens up and accepts water easily, forming a gel that does not crystallize and get stale. Unfortunately for dieters, this means that all its calories are quickly available. The ease with which amylopectin can be digested into glucose is why white bread has a higher glycemic index than pure sugar. Only half of the sugar is glucose, while all of the amylopectin is.

The image below is a small fragment of an amylopectin molecule, showing a single branch point, with each glucose molecule shown in a different color to make it easier to pick them out.

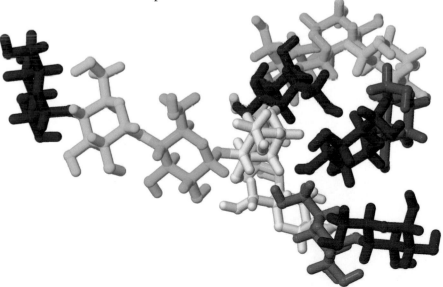

Below is another fragment, this time of a helix with another helix branching from the center. Here we have colored the many glucose units with gradually changing colors, since there are so many.

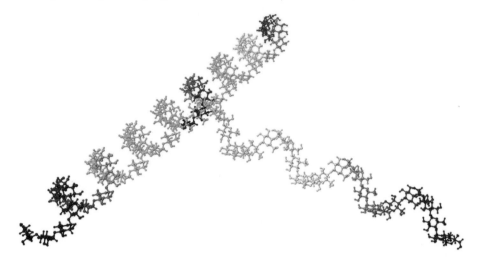

Amylose molecules are large, but amylopectin molecules are huge, with up to a couple million glucose molecules in each amylopectin molecule. As the molecule gets larger, the branching and spiraling make it fluffy and globular, as the fragment below shows:

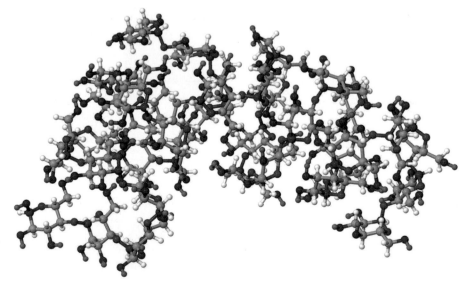

There are three types of amylopectin. Type A is found in most cereal grains, such as wheat. Wheat is generally about 26 percent amylose and 74 percent amylopectin A. Type B is found in bananas and potatoes. It is less easily digestible than type A, so more of it ends up in the large intestine to feed the good bacteria. It releases glucose into the blood more slowly than type A, producing less insulin response. The least digestible form, type C, is found in beans and legumes. More of it makes its way to the large intestine, where the bacteria ferment it into the gas those foods are famous for.

Because amylopectin is so quickly digested, it raises insulin levels quickly. Anything that does this eventually leads to insulin resistance, a key component of metabolic syndrome, type 2 diabetes, and obesity (Byrnes, Miller, and Denyer 1995).

The relative amounts of amylose and amylopectin differ in food from different plants. Even in the same plant—wheat, oats, potato, etc.—there are different cultivars with different starch ratios. One type of corn (waxy maize) has been developed to be 100 percent amylopectin, since that starch is preferred for many industrial uses.

Starch is stored in plants in tiny granules of semicrystallized amylopectin and amylose. In older milling processes, the kernels of grain would be ground to flour that kept these granules intact for the most part. In modern milling machines, the flour is ground into superfine particles where the starch molecules can more easily react with water to form gels. This fine flour is very quickly turned into glucose during digestion. This means the starch will quickly raise blood glucose levels, causing insulin levels to spike, so the glucose is stored in the body as fat.

Starches can be divided into three groups based on the speed of digestion. The quickly digested, high-amylopectin A starches, such as those in wheat flour, are in one group. More slowly digested starches such as beans and legumes form a second group. The third group is the resistant starches such as those found in cold cooked potato stored long enough for the amylose to crystallize. To control weight and avoid type 2 diabetes, reducing or eliminating the first type is a good idea.

If your starch source has at least 50 percent amylose (but preferably 70 percent or more), both blood sugar and insulin levels will be significantly

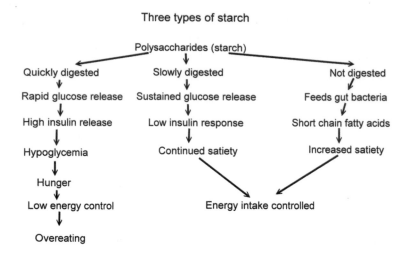

Three types of starch

reduced compared to starches with lower amylose (high amylopectin) ratios (Behall and Hallfrisch 2002). This effect is independent of whether some of the amylose has become resistant or not; in other words, it is digested but more slowly, so glucose and insulin do not rise as much (Granfeldt, Drews, and Björck 1995).

ALCOHOL

Alcohol relates to obesity in several ways. Alcohol affects the same opioid receptors in the brain that eating rich food does. People who are more prone to alcoholism, because of variations in the genes that code for opioid receptors, are more prone to bingeing on sweet and fatty foods. They are more likely to become addicted to either of these, as well as to opiate drugs, cocaine, and nicotine.

Alcoholic drinks are empty calories for the most part. A 12-ounce beer can contain anywhere from 55 calories for the lightest of light beers to well over 300 calories. A glass of wine is around 125 calories. Those calories are not coming from protein or fat. They are from alcohol itself and from sugars and starches. They do not satisfy hunger. In fact, like sugar, they actually increase hunger (Suter and Tremblay 2008). Moreover, people do not compensate for their alcohol calories by reducing food intake (Westerterp-Plantenga and Verwegen 1999). The alcohol calories are on top of their normal dietary calories, just as the calories from soft drinks are. Drinking alcohol makes us eat faster at a meal and spend more time eating (Jéquier 1999).

Alcohol has 7 calories per gram. That is almost twice as much as sugar, starch, or protein has, and almost as much as the 9 calories per gram that fat has. The body cannot store alcohol like it can store glucose or fat, so it has to burn the alcohol immediately to get rid of it. Any other calories you consume in the meantime go straight to your waistline.

At some point in alcohol overconsumption, the amount of alcohol the liver can burn through exceeds its capacity, and the toxin must be removed in other ways, such as converting it to the fatty acid acetate. By this time, however, you are definitely drunk. And while you may excrete some amount of the fatty acid in your urine, your body can absorb acetate and keep those calories as fuel. In alcoholics and other chronic drinkers, a separate pathway to detoxification is eventually triggered in order to get rid of the alcohol, and some part of it is directly oxidized, producing only heat. This occurs only when the liver cannot burn the incoming alcohol fast enough, so most of the calories have already displaced food calories, which went into fat instead of being burned (Lieber 1987).

Alcohol inhibits fat burning. The fat that is not burned is preferentially deposited in the abdomen (Suter and Tremblay 2008).

Heavy drinkers add anywhere from 1,000 to 3,000 extra calories to their diet every day. Compare that to a normal day's calories of 2,000 to 2,500. Heavy drinking also makes it difficult to get exercise. When it is difficult to walk, you are unlikely to be playing a lot of tennis.

Alcohol negatively affects sleep and growth hormone levels (Prinz et al. 2009). Disturbed sleep causes stress, which causes obesity. Human growth hormone (HGH) increases muscle mass and decreases fat mass. When it is in low supply, fat mass rises, at the expense of muscle.

About 15 percent of the calories in alcohol go into thermogenesis—creating body heat. This compares with 25 percent for protein calories, 8 percent for carbohydrate calories, and 3 percent for fat calories (Jéquier 1999). The remaining 85 percent of the calories from alcohol are available to the body as energy and are used before energy from food sources, leaving that for fat storage.

DIETARY FIBER

Fiber in the diet has many effects relevant to weight control. It is filling, yet has no calories. It absorbs and retains water, aiding satiety. It slows the emptying of the stomach. It decreases hunger without adding calories.

An additional 14 grams of fiber per day is associated with a 10 percent decrease in energy intake and a weight loss of over a pound a month (Howarth, Saltzman, and Roberts 2001). Moreover, these effects are greater in obese individuals than in lean ones.

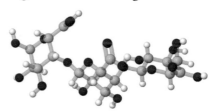

A small fragment of the fiber pectin.

Since the average American's fiber intake is only 15 grams per day, this is effectively doubling the average intake, to reach the AMA-recommended 25 to 30 grams per day.

In a study of female nurses, those with the highest fiber intake had a 49 percent lower risk of weight gain than those with the lowest intake (Liu et al. 2003). Fiber intake correlates with lower body mass index (BMI) no matter how much fat is in the diet (Slavin 2005).

Foods high in fiber take longer to eat. Consider the 250 calories in a vanilla latte compared to the same 250 calories in 10 carrots. Not only is the latte finished in less time than a single carrot, but 10 carrots is a meal

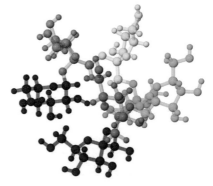

A small fragment of the fiber inulin.

that leaves you quite full, whereas the latte just makes you hungrier. The time it takes to eat your calories matters quite a bit. It takes time for your body to register that you have eaten enough. If you eat your whole meal in 20 minutes or less, you won't have time for the satiety signals to develop or reach your brain. You will eat until you are stuffed, instead of eating until you are no longer hungry. The difference can be several hundred calories per meal, and that adds up to pounds.

Fiber reduces the rate at which the intestines absorb carbohydrates. This reduces the insulin response, since it is sensitive to the *rate* of glucose absorption more than to the *amount* of glucose.

Fiber speeds the food through the small intestine. This allows the food to get to the large intestine sooner. It is in the large intestine that the satiety hormone PYY is produced, telling us to stop eating. Fiber tells us we are full sooner, so we don't keep eating more calories.

Fiber attaches to free fatty acids, so they don't get absorbed as easily in the small intestine. When they reach the large intestine, the beneficial bacteria there chop them up into short-chain fatty acids, which act as a signal to suppress insulin production in the pancreas. Without the fiber, more of the fatty acids remain as long chain fatty acids, which have the opposite effect. Long-chain fatty acids increase insulin production. They also make muscle cells less sensitive to insulin.

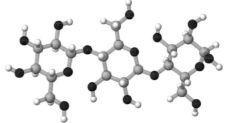

Besides weight management, high levels of dietary fiber are associated with lower risk of heart disease, stroke, high blood pressure, diabetes, and many gastrointestinal diseases. Fiber reduces cholesterol in the blood. Soluble

A small fragment of the fiber beta-glucan.

fiber improves insulin sensitivity and glycemia. Fiber improves immune function by feeding beneficial bacteria in the large intestine.

In a study comparing a high-fiber diet (50 grams per day) to one with 24 grams in diabetic patients, the high-fiber diet reduced cholesterol absorption by 10 percent, improved blood glucose and insulin levels, and reduced glucose levels in the urine (Chandalia et al. 2000).

Dietary fiber comes in two forms: soluble and insoluble. The soluble form is what makes gels and raises the viscosity of the stomach contents. Insoluble fiber speeds digestion and causes some of the energy intake from the diet to end up being undigested. The soluble form is digested by gut bacteria producing short-chain fatty acids, which are then absorbed by the gut. This leads to higher amounts of energy being absorbed. Of course, replacing fat and carbohydrate with fiber, even soluble fiber, results in reduced energy intake. The soluble fibers are what reduce cholesterol and triglycerides in the blood.

There are two mechanisms by which soluble fiber reduces cholesterol in the blood. Fiber binds to bile acids in the intestines. Bile acids are derived from cholesterol, so when the fiber binds to them and carries them out of the body, new bile acids must be formed from cholesterol in the blood.

Bile acids are needed in order to digest fats and fat-soluble vitamins. They act like detergent to break up fat globules into tiny bits that can be absorbed.

The second way fiber helps to reduce cholesterol in the blood happens when the fiber feeds the beneficial bacteria in the large intestine.

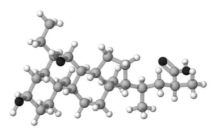

Bile acid acts like a detergent to digest fats.

These bacteria ferment the fiber and produce short-chain fatty acids: acetate, propionate, and butyrate. These short-chain fatty acids act like hormones in the body and change the rate at which cholesterol is synthesized in the liver and small intestine. Acetate increases the rate of cholesterol synthesis, while propionate reduces it. Many of the butyrate-producing bacteria in the gut eat acetate as their fuel, resulting in less acetate getting into the bloodstream.

Butyrate is the principal fuel for the cells lining the large intestine (colonocytes). Without it, they die. Lack of sufficient butyrate is implicated in colon cancer. Butyrate is also an inflammation moderator, with anti-inflammatory effects, which are important in weight management because inflammation is a key aspect of obesity, metabolic syndrome, diabetes, and coronary heart disease.

FATS AND PROTEIN

DIETARY FAT

For years, the accepted medical wisdom was that eating fat made you fat. It seemed to make sense. Dietary fat has more calories per gram than protein or carbohydrates, the other two macronutrients. We gave up fat and replaced it with sugar, and we got fat on low-fat diets.

Worse still, the kinds of fats we started to eat were the worst types. We replaced lard and butter with hydrogenated vegetable oils and cheap seed oils high in inflammatory omega-6 fatty acids and low in healthy omega-3 fatty acids. We considered saturated fats to be the worst kind, because we thought they contained only calories, unlike unsaturated fats, which the body requires to make important tissues and molecules.

We have since learned that dietary fat is important for health and weight management. Grass-fed butter has important omega-3 fatty acids. Saturated fats, especially those containing medium-chain triglycerides such as butter and coconut oil, are important for brain function, promoting healthy gut bacteria, controlling inflammation, and raising your metabolic rate.

Dietary fat relieves hunger, unlike the sugar and processed starches that replaced it. Consequently, we ate more, demanded larger servings at restaurants, and bought bigger dinner plates.

To understand how your body processes and uses dietary fat, it helps to look into the chemistry of it. After all, the words we use, such as *triglycerides*, *fatty acids*, *omega-3*, *omega-6*, *saturated*, and *unsaturated*, are all words borrowed from chemistry.

Fats and oils are made of a small molecule called glycerol, to which are attached three molecules called fatty acids. This is why we call them triglycerides.

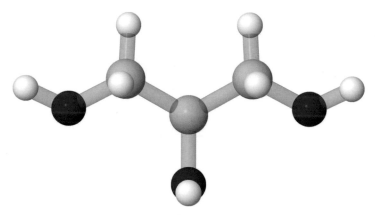

The glycerol molecule is shown above. The carbon atoms are gray, the oxygen atoms are red, and the hydrogen atoms are white.

Fatty acids are chains of carbon atoms with an oxygen atom at one end, which binds to the glycerol, and hydrogen atoms are attached all around. The simplest fatty acid is formic acid, which has a single carbon atom.

Other simple fatty acids (called short-chain fatty acids), containing two, three, and four carbon atoms, are acetic acid, propionic acid, and butyric acid, respectively:

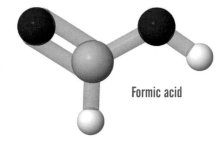

Formic acid

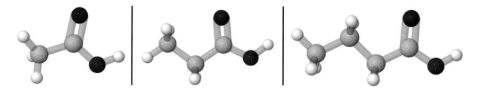

When three fatty acids are attached to a glycerol molecule, we get a triglyceride, which is what we know as an oil or a fat. The simplest triglyceride would be glycerol triformate. It is made from three formic acid molecules attached to the glycerol molecule (the three carbons in the center).

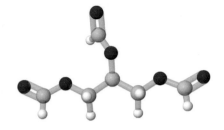

Glycerol triformate

Most of the oils and fats you might be familiar with contain much longer fatty acids, with chains of carbon that are eight atoms long or longer, up to as many as 38. The longer the chain, the more likely the resulting molecule is to be solid.

Saturated fatty acids are those with only single bonds between the carbons. They are called saturated because they have as many hydrogen atoms as will fit on the molecule. Single bonds allow the molecule to rotate and bend easily, and saturated fats can thus pack together easily, so they can cluster closely and become a solid. This ability to pack tightly makes them useful as a way to store calories in an animal or in a seed.

A common saturated fatty acid found in animal fats and cocoa butter is stearic acid. It has 18 carbon atoms in its chain:

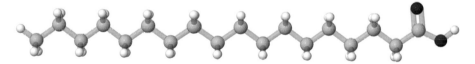

Unsaturated fatty acids have at least one double bond in the chain. A double bond has two carbon atoms that are joined by two bonds. Since each carbon atom can have only four bonds, those two carbons can have only one hydrogen atom each. Double bonds can't rotate—they are like a door with two hinges and can only swing side to side. This means unsaturated fats don't pack as easily as saturated fats do, so they are more likely to be liquids. An example of an unsaturated fat is oleic acid, on the right.

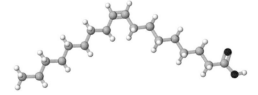

You can see the double bond in the center. Oleic acid is a mono-unsaturated fatty acid. This means it has only one double bond. Fatty acids with more than one double bond are called polyunsaturated fatty acids.

A triglyceride can have any of the fatty acid types attached to its glycerol backbone. As an example, adding one saturated fatty acid (palmitic acid), one monounsaturated acid (oleic acid), and one polyunsaturated acid (alphalinolenic acid) gives us 1-oleoyl-2-palmitoyl-3-a-linolenoyl-glycerol:

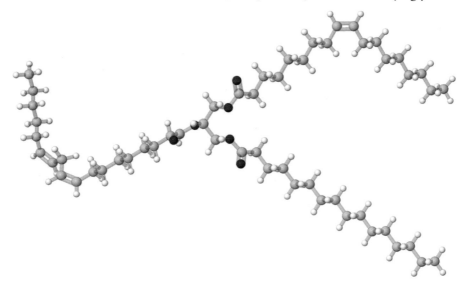

Unsaturated fatty acids are categorized by where the double bond is located on the carbon chain. We number the carbons from the end opposite from the glycerol. We use the Greek letter omega to refer to the last carbon (omega is the last letter in the Greek alphabet). A fatty acid that has a double bond between the third and fourth to the last carbons is an omega-3 fatty acid. If the double bond is between the sixth and seventh carbons from the end, it is an omega-6 fatty acid. Oleic acid is an omega-9 fatty acid. Its double bond is between the carbon atoms that are ninth and tenth from the end.

The *trans* fatty acids are those where the hydrogen atoms attached to the double bond carbons are on opposite sides of the carbons. If the hydrogen atoms are on the same side, as they are in oleic acid, we say they are *cis* fatty acids. Converting oleic acid from the cis form to the trans form gives us elaidic acid:

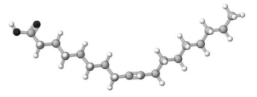

Trans fatty acids are formed when fats are overheated, as they are when natural oils are artificially hydrogenated (adding hydrogen to some of the double bonds to make them more saturated). This used to be a common way to make cheap seed oils into solids more like lard or butter. The result was margarine and shortening, cheaper alternatives to the animal fats.

Trans fats have been found to increase the bad form of cholesterol (LDL) more than the saturated fats they were designed to replace. Trans fats also increase inflammation, which leads to heart disease, obesity, and diabetes. They now must be listed in ingredients labels so they can be avoided, and as a result, their use has fallen considerably.

Not all trans fats are harmful, however. Several natural trans fats are actually beneficial. Dairy fat contains trans-palmitoleic acid, which has been found to be protective against type 2 diabetes, and thus heart disease. Test subjects with high blood levels of this trans fatty acid were found to be 60 percent less likely to develop diabetes than those with the lowest levels in their blood (Mozaffarian et al. 2010; Mozaffarian et al. 2013).

While saturated fats can raise cholesterol in the short term (a few weeks), long-term studies have shown the association to be weak or nonexistent. Many saturated fats, such as stearic acid and medium-chain triglycerides, are actually beneficial to heart health. Large studies have found that saturated fat intake has no association with heart disease, and that saturated fat is actually beneficial in preventing stroke (Yamagishi et al. 2010).

The types of fat in the diet affect our health in many ways. Some fatty acids (arachidonic acid [AA]) cause inflammation, which causes obesity and diabetes, while others (eicosapentaenoic acid [EPA] and docosahexaenoic acid [DHA]) reduce inflammation. Some fats (medium-chain triglycerides) increase our metabolic rate, causing us to burn more calories.

Some fatty acids cannot be manufactured in the body, so we must get them from food, as they are important building blocks of cell membranes and other biological structures. These are called essential fatty acids. Humans lack the enzymes needed to insert double bonds at the omega-3 and omega-6 carbons, so the essential fatty acids are all omega-3 or omega-6 fats. The two shortest essential fatty acids are the omega-6 fat linoleic acid (LA) and the omega-3 fat alpha-linolenic acid (ALA). All the longer essential fatty acids can be made from these by enzymes that add carbons or add double bonds, but it is obviously more efficient to get them premade from the diet. The importance of DHA and EHA, and the inefficiency of creating them from ALA, make these two omega-3 fatty acids important dietary components. They are found mostly in seafood such as oily fish, but the best source is krill, a crustacean. Krill oil has the omega-3 fatty acids bound to phospholipids that make them more available to the body. Krill oil also contains astaxanthin, a flavonoid that also increases the bioavailability. The astaxanthin in krill oil is one of the more potent antioxidants (better than vitamin E and beta-carotene), and can help prevent the fatty acids from oxidizing and consequently becoming rancid (Naguib 2000). Astaxanthin is 48 times better than fish oil and 34 times better than the coenzyme Q10 as an antioxidant (Massrieh 2008).

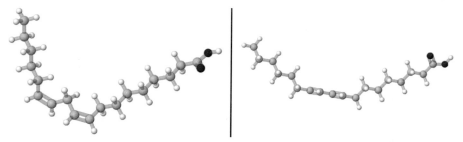

The two fats shown above are linoleic acid (LA) and conjugated linoleic acid (CLA). They are the same except that in the second one the double bond has moved from the omega-6 carbon to the omega-7 carbon (the seventh from the last instead of the sixth from the last).

Linoleic acid is an omega-6 fatty acid, but not one of the inflammatory ones. It is benign as a dietary food source, although high insulin levels cause enzymes to convert it into inflammatory arachidonic acid. Conjugated linoleic acid has many health benefits: fighting cancer, obesity, atherosclerosis, asthma, high blood pressure, high cholesterol, insulin resistance, and osteoporosis, and improving immune function (Bhatta-charya et al. 2006). Pertinent to obesity, the combination of CLA and exercise lowers fat and increases lean body mass (Bhattacharya et al. 2005).

Bacteria in the gut convert linoleic acid into conjugated linoleic acid. Cows fed on grass produce milk with substantially more CLA than grain-fed cows. The same goes for beef cattle—those fed on grass have more CLA.

Dietary Fat and Obesity

Part of the obesity problem is that we stopped eating enough fat. Fat in the diet reduces hunger. We feel full. Satisfied. When we replaced fat with sugar and processed carbohydrates, we started eating more at each meal, and more between meals, because calorie for calorie, those foods don't satisfy hunger as well as fat does. Some carbohydrates, like fructose, actually make us hungrier the more we eat them.

We also were eating the wrong kinds of fat. Trans fat and cheap seed oils contain inflammatory fatty acids that cause insulin resistance. Inflammation of the brain causes leptin resistance. Leptin is the hormone fat cells produce to tell the brain we have had enough to eat. Too much omega-6 fat (particularly arachidonic acid) in relation to omega-3 fats in the modern industrial diet leads to more inflammation and obesity.

Omega-3 fats reduce inflammation. Medium-chain triglycerides raise our metabolic rate, so we burn calories faster, even when not exercising.

Fat cells that are not completely full are healthy. They can pull glucose from the blood to make glycerol and pull fatty acids from the blood to attach to the glycerol to make triglycerides for storage. Stored triglycerides are not just a source of energy in lean times. They are a way of removing harmful glucose and inflammatory fatty acids from the blood. Insulin is a signal to them to do this.

When the blood contains high levels of the inflammatory fatty acid known as arachidonic acid (AA), the healthy fat cells remove it from the blood.

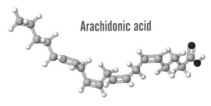

Arachidonic acid

But when the levels of AA in the fat cell become too high, the body defends itself by generating new fat cells. The signal to do this comes from breakdown products of arachidonic acid. The new fat cells help to remove fat from the blood so it does not get stored in muscles and other organs where it can be toxic. This is called lipotoxicity (Sears and Ricordi 2010).

A problem arises when levels of arachidonic acid in the fat cell become too high. The AA converts into inflammatory compounds called eicosanoids, and these interfere with the fat cell's ability to respond to insulin.

The flow of glucose into the fat cell is interrupted, and it can no longer take up fatty acids from the blood. Now the fat cell cannot remove AA from the blood. Fatty acids build up in the blood and muscles, where they interfere with insulin signaling, so the muscles don't burn glucose. The fat cells that are too full now start releasing fatty acids into the blood, making the situation worse. At some point, the AA levels in the fat cell become so high the cell dies. This causes macrophages to enter the fat tissue to clean up the debris from dead fat cells. They release inflammatory molecules that cause further inflammation in the fat tissue, further reducing insulin sensitivity. These molecules also migrate to the liver, where they cause it to produce C-reactive protein. Doctors use levels of C-reactive protein as a marker for chronic inflammation.

Prostaglandin is an eicosanoid formed from arachidonic acid.

When the fat cells are busy creating inflammation, diseases of inflammation, such as diabetes and heart disease, start to show up.

About a third of obese people have genes that make their fat cells produce more of the hormone adiponectin than other people (adiponectin will be discussed in more detail in the chapter on hormones, page 89). These people are protected from inflammation and its consequences. For the rest of us, getting high levels of the omega-3 fatty acid EPA can stimulate fat cells to produce more protective adiponectin.

Adiponectin reduces insulin resistance, so the fat cells can do their job, and it stimulates the production of new healthy (empty) fat cells that are able to remove glucose and fatty acids from the blood (Sears and Ricordi 2010).

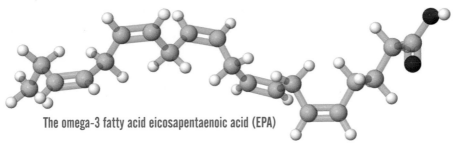

The omega-3 fatty acid eicosapentaenoic acid (EPA)

PROTEIN

A key component of recent popular diets such as Atkins and Paleo is the relatively high consumption of protein. There are several reasons why obese and overweight people benefit from extra protein. For children and lean individuals, extra protein may not be beneficial and may actually cause problems. High-protein diets are associated with impaired glucose tolerance, reduced insulin sensitivity, and increased incidence of type 2 diabetes (Chen and Yang 2015).

Eating protein causes levels of the appetite suppressant hormone glucagon to rise more than eating an equal number of calories of carbohydrate. Calorie for calorie, protein curbs hunger better than carbohydrates (Arafat et al. 2013) and fat (Keller 2011) do. The satiety effect of a high-protein diet cannot be explained by changes in the other appetite-regulating hormones ghrelin and leptin (Weigle et al. 2005). Protein may also increase the brain's sensitivity to leptin.

Protein targets satiety, thermogenesis (body heat production), energy efficiency, and body composition (fat ratio) (Keller 2011).

When we eat anything, the rate at which we burn calories for heat (thermogenesis) goes up. But for protein it increases by 20 to 30 percent, while for carbohydrate it increases by 5 to 10 percent, and for fat it increases by at most 5 percent (Keller 2011).

Eating protein while losing weight preserves muscle mass as well as bone mass. In a study of subjects who had just completed a very low-calorie

diet, one group was assigned a diet with 18 percent of calories from protein, while the other group got 15 percent of calories from protein. Those
getting more protein regained half as much weight as the control group,
and that weight was lean body mass, not fat. This was attributed to two
factors: the higher satiety of the high-protein diet, and the reduced energy
efficiency (Westerterp-Plantenga et al. 2004). Lean body mass recovery is
important, as the weight loss from the very low-calorie diet was 61 percent
from fat and 39 percent from fat-free mass (muscle, bone, water).

In another study, subjects assigned high-protein diets and low glycemic
index diets both fared better than controls at maintaining weight loss,
and the combined group (high protein and low glycemic index) did even
better (the effects were additive). All groups were allowed to eat as much
as they wished from the selected foods. There is a genetic component to
whether high-protein, low glycemic index diets are effective in people,
but two-thirds of people have the genotype that makes this diet effective
(Astrup, Raben, and Geiker 2015).

In one large-scale randomized controlled trial, the DioGenes trial, 932
obese families were studied through an eight-week, 800-calorie-per-day
diet, followed by six months of five different diets that allowed them to eat
all they wanted, while controlling for how much protein and which kind
of carbohydrate (low glycemic index vs. high glycemic index) they ate.

The group that ate a high-protein, low glycemic index diet kept the
weight loss for over a year (the period of the study), while the other groups
gradually gained the weight back (Astrup, Raben, and Geiker 2015). Of
the two diet components, the protein was more important than the low
glycemic index carbohydrates in keeping off the weight.

In that trial, fewer of the high-protein, low glycemic index group
dropped out of the trial, indicating that this diet was easier to stick to.
This group also lost 5 percent more weight than the other groups. The
low glycemic index dieters also lost 5 percent more than the high glycemic
index dieters.

The take-home lesson from the study was that subtle changes in diet
composition can have a major impact on calorie intake in real-life situations. The combination of high protein with low glycemic index had an

additive effect. Either one was good, but both together was better. That the diet was easier to stay on also had a noticeable effect.

On the downside, getting too much protein eliminates one of the benefits of weight reduction—the increase in insulin sensitivity (Smith et al. 2016). The diets studied showed that the high-protein diet reduced the loss in lean muscle mass by 45 percent—an important finding—but at the cost of no improvements in insulin sensitivity. On a lower-protein (but still low-calorie) diet, insulin sensitivity improved. The protein content of the diets was 1.5 grams per kilogram versus 0.8 grams per kilogram. For a 150-pound woman, that would be 81 grams of protein per day versus 54 grams.

Further concern about extra protein, especially animal protein, centers around the timing of its consumption in children. Getting too much dairy protein at the age of 1 year is associated with extra body fat at year 7, and body fat percentage at year 7 correlates well with body fat percentage later in life. Getting too much animal protein in years 5 and 6 is associated with obesity later in life (Günther et al. 2007). Vegetable protein did not have this effect on future weight gain.

Infant formula has 50 to 80 percent more protein than breast milk, and there is a higher incidence of later obesity in formula-fed babies than in breast-fed babies. Moreover, in transitioning from milk to solid foods, there is, in many populations, a rapid increase in protein intake, exceeding protein recommendations by up to five times as much.

Cow's milk, but not meat or vegetable protein, may stimulate insulin and insulin-like growth factor (IGF-1), two hormones that act throughout the body to change both body composition and how we react to foods. Specifically, these two hormones affect the differentiation and proliferation of preadipocyte cells—cells that go on to become either beige or white fat cells. The timing of this protein intake seems to be important, with the age of 12 months being critical (Günther et al. 2007). Dairy protein consumption later in life had no effect on future weight gain. There are indications that the bovine whey protein is what stimulates insulin secretion, so presumably cheeses would have a lesser effect.

The age of 12 months is when children transition from milk-only diets to those containing solid food. An excess of dairy protein at this time seems to lead to later weight gain. Other animal protein and vegetable protein don't have any negative associations with later obesity.

In an intervention study, replacing formula containing 11.7 percent of calories from protein with formula containing only 7.1 percent of calories from protein led to growth at age 2 more like that of breast-fed children. For comparison, human breast milk contains only 5 to 6 percent of calories from protein (Koletzko et al. 2016).

PHYSIOLOGY AND PSYCHOLOGY

SLEEP

Sleep disorders and obesity go hand in hand. Sleep apnea (pauses in breathing or shallow breaths while you sleep) is caused by insulin resistance associated with abdominal fat (Vgontzas, Bixler, and Chrousos 2003). But waking up frequently in the middle of the night from lack of oxygen is not the only obesity-related sleep disorder. A disorder known as "excessive daytime sleepiness" is caused by tumor necrosis factor alpha and interleukin-6, two chemical messengers produced in excess by the abdominal fat tissue in obese people. These molecules act on the brain in the areas where sleep and wakefulness are controlled, in addition to their many other effects throughout the body. The levels of these two chemical messengers rise with increases in body mass index (BMI). The fatter we are, the more daytime sleepiness happens.

People with other symptoms of insulin resistance are more likely to have sleep disorders than people with normal insulin sensitivity. For example, ovarian cysts and diabetes are associated with sleep disorders, likely by the same mechanisms.

Getting rid of abdominal fat leads to improvements in sleep. The most dramatic examples of this happen with the most drastic interventions.

Gastric bypass surgery leading to dramatic weight loss can reduce or eliminate sleep apnea.

Men, with their greater propensity to accumulate abdominal fat (beer belly) than women, who tend to distribute the fat more than men, are also more likely to have sleep apnea (Vgontzas, Bixler, and Chrousos 2003). In women, sleep apnea tends to occur after menopause, when fat distribution becomes more male-like.

Sleep apnea caused by obesity and insulin insensitivity causes stress due to lack of sleep. This stress causes the release of cortisol (a stress hormone), which further increases obesity and insulin resistance.

Lack of sleep for any reason, such as night shift work, has a similar result. It causes stress, which releases stress hormones, which cause obesity and insulin resistance.

Sleep is an important modulator of metabolic function (how much we eat and how fast we burn it) and endocrine function (hormone production). It also modulates the cardiovascular system, with effects on blood pressure. Sleep deprivation studies have shown that 60 hours without sleep reduces insulin sensitivity (VanHelder, Symons, and Radomski 1993). In a study where healthy young volunteers were allowed only four hours of sleep per night for six days, glucose tolerance was significantly lowered (Spiegel, Leproult, and Cauter 1999). Self-reported "short sleepers" also had impaired glucose metabolism compared to control groups that got enough sleep.

Lack of sleep makes you hungry. Levels of ghrelin (the hunger-stimulating hormone produced by the stomach) go up. Levels of leptin (the satiety hormone produced by fat cells) go down (Taheri et al. 2004). This double whammy causes us to eat more (Spiegel et al. 2005). Conversely, a shortage of food leads to decreased or disturbed sleep. The same molecules that cause wakefulness by acting on the brain's hypothalamus also stimulate food intake. We get hungry. On the other side of the equation, as we start to sleep, the levels of the satiety hormone leptin start to rise. It is easier to sleep when we don't feel hungry.

Most mammals sleep in short bouts, whether they sleep at night or during the day. Humans, however, sleep in 7- to 9-hour consolidated periods. This means that there is an extended period of fasting, when a sleeping human is not eating. During sleep, blood sugar levels stay fairly level. In contrast, during fasting while awake, glucose in the blood falls continually. The body has a number of mechanisms for keeping blood sugar levels constant during the sleeping fast.

While a person sleeps, cortisol levels fall. This would be expected, since sleep is seldom stressful. This period of low cortisol levels increases insulin sensitivity. Conversely, disturbed sleep is stressful, and leads to insulin resistance and obesity.

Human growth hormone levels, especially in men, are stimulated during sleep. In men, 60 to 70 percent of all growth hormone secretion happens during sleep. Human growth hormone (HGH) is produced in the pituitary gland at the base of the brain. It fuels growth and development in children but is still produced in adults, where it has some effect on reducing fat and increasing muscle mass. Obese adults have lower human growth hormone levels in the blood than lean adults do. Older adults have less HGH than younger adults. We will discuss HGH in greater detail later (page 107). For now, it is sufficient to mention that most of it is secreted while a person is asleep, and that it increases muscle mass and decreases fat mass.

The hormone melatonin controls the sleep-wake cycle.

Melatonin levels are controlled in part by the amount and type of light we're in. Blue light, such as daylight sky, or light from television and computer screens, inhibits melatonin and prevents sleep. Older adults make smaller amounts of melatonin, leading to less sleep. Melatonin stimulates HGH production, so less melatonin

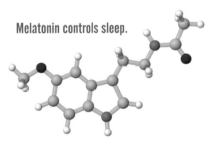

Melatonin controls sleep.

means less sleep and less HGH. Seasonal affective disorder, where low light levels during the temperate winter cause depression, is sometimes treated with over-the-counter melatonin supplements. Melatonin supplements are also used for night shift workers and for treating jet lag and insomnia.

Ghrelin (an appetite-stimulating hormone) increases from midnight to dawn in thin people but not in obese people. This indicates that obese people have a flaw in their circadian rhythms, the same rhythms that make us sleepy at night.

In the past 30 years, we have seen obesity rise to epidemic levels. Over the same time period, the average duration of nightly sleep has fallen from eight to nine hours to only seven hours per night (Bose, Oliván, and Laferrère 2009). Nowadays, 30 percent of all adults in the United States sleep less than six hours per night. The results are increased BMI and insulin resistance.

STRESS

When there is a challenge to the natural homeostasis of an organism, we call that "stress." The reaction to stress is to produce a physiological reaction to regain equilibrium. This takes the form of behavioral and physical adaptations, such as increasing cognition (your brain gets sharper), increasing the production of glucose by the liver, through the breakdown of both muscle and fat, and inhibiting reproduction (Bose, Oliván, and Laferrère 2009).

There are two systems involved in the stress response: the autonomic nervous system (ANS) and the hypothalamic-pituitary-adrenal (HPA) axis. They work together to activate one of two responses: the "fight-or-flight" reaction or the "defeat response" reaction (where the animal loses the fight). The latter is the predominant stress response to modern life and is primarily handled by the HPA system. The HPA system is the stress response system associated with upper-body obesity, metabolic syndrome, and depression.

The hypothalamus is the organ in the brain that releases corticotropin-releasing hormone (CRH) and stimulates the pituitary gland to release adrenocorticotropin (ACTH). ACTH tells the adrenal glands to release the stress hormone cortisol and other glucocorticoid hormones.

Glucocorticoid hormones interact with many systems in the body, including those governing hunger and satiety. They interact with insulin and the hormone neuropeptide Y to increase hunger and food-seeking behavior. People taking the artificial glucocorticoid drug prednisone

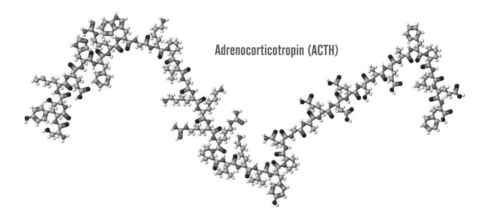

Adrenocorticotropin (ACTH)

note appetite stimulation and eat more. Long-term use of such drugs can produce Cushing's syndrome, characterized by substantial upper-body obesity.

Mental stress has been shown to cause an increase in blood levels of interleukin-6, a key inflammatory molecule (Dandona, Aljada, and Bandyopadhyay 2004). Stress can induce inflammation, which induces obesity. Stress interferes with glucose homeostasis in diabetics, exacerbating their condition.

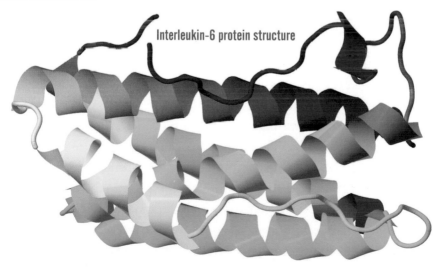

Interleukin-6 protein structure

Stress hormones such as cortisol, epinephrine (adrenaline), and norepinephrine have all been linked to obesity, and chronic stress affects insulin and glucose metabolism through these hormones and through their receptors. Epinephrine in particular lowers leptin levels, which increases hunger.

There is a difference in the brain chemistry between acute stress and chronic stress. Acute stress is the reaction to an isolated stressful event, such as an accident, getting a bad grade, sleep deprivation, getting fired, moving to a new house, etc. In acute stress, the brain sends signals to the adrenal glands, which produce hormones called glucocorticoids, which increase attention and arousal, among many other effects (Dallman et al. 2003). But then comes an inhibiting signal that quenches the glucocorticoids after about 18 hours. This type of negative feedback is common in most control systems, and it keeps the system balanced.

Chronic stress switches the brain into a different mode. After about 24 hours of stress, the negative feedback signal diminishes, and the brain enables the "chronic stress-response network." This network is associated with a variety of coping mechanisms. These include salience, where attention is focused on the stressor or other environmental cues that may or may not be related (Dallman et al. 2003). Things start to stand out from the background. Things that are rewarding become much more important, often to the point of becoming compulsive.

With chronic stress, the levels of glucocorticoids stay high, since the negative feedback control is reduced. Glucocorticoids cause muscle to be broken down so that its component amino acids can be used in the liver to make glucose. They inhibit growth hormone, whose normal effect is to prevent this muscle wastage and to mobilize fat stores instead. This reduces calorie efficiency. Glucocorticoids cause us to desire more sugar and fat and other rewarding foods, and at the same time shift the fat storage from under the skin (subcutaneous fat) to abdominal fat.

Negative feedback is crucial to any control system. In the chronic stress-response network, the negative feedback comes from two sources that are different from that in acute stress. One is the reward system in the brain. When we are rewarded, glucocorticoid levels drop. This is why we call certain highly palatable foods "comfort foods." They actually do comfort us in situations where the stress levels have been high for a long time.

The second feedback signal comes from abdominal fat deposits. As we compulsively feed on comfort foods, and the glucocorticoids move fat from under the skin to the central abdominal area, the abdominal (visceral) fat increases. This visceral fat sends signals to the brain to reduce glucocorticoid levels. The body is saying that we have enough stored fat to get us through the stressful situation. By the time this happens, we have dangerous levels of abdominal fat, with its association with diabetes, stroke, hypertension, and other aspects of metabolic syndrome.

The compulsions caused by elevated levels of glucocorticoids lead to eating disorders such as bingeing and night-eating syndrome, in which most of the day's calories are eaten at night. People with these disorders report being chronically stressed (Dallman et al. 2003). The foods they binge on are typically high in sugar and fat. These people are also, not surprisingly, usually obese. The glucocorticoid levels in these people are only slightly elevated.

In people with anorexia nervosa, on the other hand, the levels of the stress hormone cortisol (a glucocorticoid) are very high and insulin levels are very low, yet these people still have higher levels of abdominal fat than subcutaneous fat.

In both groups, rates of depression are very high. Related to this is a common side effect of antidepressants, increasing obesity.

The difference between binge eaters and anorexics is how they try to cope with the stress. Binge eaters are trying to reduce the cortisol levels through the reward system, while anorexics are using the seeking or escape mechanisms. Glucocorticoids are what make rats run on wheels. They induce seeking (what the rats are running toward) and escape (what they are running from). When we remove their adrenal glands, the rats stop using the running wheel.

Self-medicating stress-induced depression with a pint of Ben & Jerry's ice cream is not a healthy long-term strategy. Take a tip from rats and instead go for a walk or a jog. Stress-relieving activities such as yoga, meditation, and exercise can be done for much longer periods, with much better outcomes.

DEPRESSION

Ten percent of the population suffers from depression, and 35 percent of the population suffers from obesity, so it is to be expected that the two conditions will be found together. However, depression can cause obesity, and obesity can cause depression, so the incidence of each is much more than merely statistical (Stunkard, Faith, and Allison 2003).

Obesity can cause low self-esteem and social isolation, which are known contributors to depression. Obese people are also prone to inflammation, which results in joint pain, diabetes, and high blood pressure, all of which are linked to depression (Luppino et al. 2010).

People with depression are more likely to become obese. Depression stimulates weight accumulation (Dandona, Aljada, and Bandyopadhyay 2004). People who are depressed make poor food choices, are more likely to overeat and less likely to exercise, and become more sedentary. Low levels of the hormone serotonin—a major cause of depression—cause people to self-medicate by eating rewarding comfort foods.

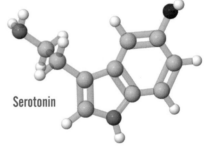

Serotonin

Forty-three percent of adults with depression are obese, compared to only 33 percent of adults without depression (Pratt and Brody 2014). People with bipolar disorder or recurrent depression with low levels of cortisol suffer from metabolic syndrome, obesity, and high cholesterol more than other people do (Maripuu et al. 2016).

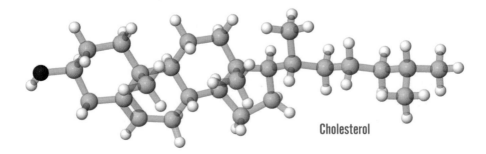

Cholesterol

Depression and obesity are both exacerbated by low social class and lack of physical activity, making the coincidence of both diseases more likely. Bipolar disorder and recurrent depression are associated with a 10- to 15-year reduction in life expectancy. Cardiovascular disease is a strong contributing factor (Maripuu et al. 2016).

As obese depressed people lose weight, many also lose their depression. A year after bariatric surgery to lose 77 percent of body weight, patients also experienced an 18 percent reduction in symptoms of depression. On the other hand, the more severely depressed someone is, the more likely he or she is to be or become obese (Pratt and Brody 2014).

The link between depression and obesity seems to be tied to sex, race, and age. Men under 60 show no association between depression and obesity, while those over 60 do, as do women of all ages (Pratt and Brody 2014). Non-Hispanic black and Mexican American men and women show no association between the two diseases. In patients who were over 20 years old, being overweight was associated with depression, but this was not the case with younger patients (Luppino et al. 2010). The longer patients suffered from depression, the more likely they were to become obese. Obese subjects had a 55 percent increased risk of depression, and depressed subjects had a 58 percent increased risk of becoming obese (Luppino et al. 2010).

Inflammatory hormones called cytokines can cause both obesity and depression. Hormones associated with inflammation affect the brain, causing sadness, fatigue, altered sleep patterns, and social withdrawal. The same hormones (see the section on inflammation, page 167) also cause obesity. Treating obesity by removing inflammatory omega-6 fatty acids and starches from the diet, and

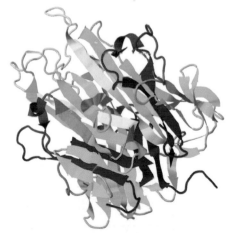

Inflammatory cytokine tumor necrosis factor alpha (TNF-alpha)

adding anti-inflammatory polyphenols (the molecules that give vegetables their colors), may also help to treat depression.

The two problems share other aspects as well. Depression and obesity can both be caused by stress. Some symptoms of depression, such as binge eating, can contribute to obesity. People who succumb to temptations while on a diet can be unduly hard on themselves, causing further stress. Treating both problems simultaneously, with the help of health professionals, usually works better than treating either condition alone.

AUTOMATIC EATING

Automatic eating is eating without thinking about whether you need to eat. You eat because the food is there.

It may be that your plate is too full, and that you or your server was not paying attention to proper serving sizes (no portion control). You clean the plate because it's there, not because you are hungry.

At dinner, you might be in a stimulating conversation with someone, and the table still has food on it. You serve yourself more food without thinking about it. You could have stopped eating a while ago, but the food was just there. To avoid this, fill the plates in the kitchen with reasonable portions and bring the plates to the table. If diners really need a second helping, they can get up and go to the kitchen. But most second helpings aren't something people think about. They eat because the food is there, not because they are still hungry.

At a movie, sitting there with a huge tub of buttered popcorn, you keep reaching in and stuffing more into your mouth, but if someone had taken the bucket away half an hour ago, you would not have noticed or minded. Again, portion control could have helped here. Get a smaller bag of popcorn.

At a party, there is food everywhere, and it is all delicious high-fat, high-sugar, high-salt goodies that feed your addiction. You can't just eat a little and stop. Your brain forgets that you said to stop, and the eating goes on automatically, sometimes to the point of bingeing. If you are dieting, tell everyone you have already eaten your daily allowance, and carry a glass of water around to sip whenever you are tempted to grab a bite. People will delight in your misery and help you by pointing out that you said you weren't going to eat that cookie you have in your hand. Let them help.

You pass by the refrigerator on your way out and can't help but look inside. You aren't hungry; it has just become a habit. Put the carrots and celery in the front so you see them every time you open the fridge. If you don't feel like eating a carrot, you aren't really hungry. You are just looking to feed your addiction to addictive foods.

On the counter is a box of cookies or a bag of chips. Throw those away, and either leave the counter empty or replace the addictive foods with a bowl of apples or some nuts. Again, if the apple doesn't look like what you want to eat right now, you aren't really hungry. If you find that you have added raisins and M&M's to your bowl of nuts, and now you always grab a handful when you pass by, then you have replaced something good with something addictive. Don't do that.

EATING FOR PLEASURE AND REWARD

Processed foods are *designed* to make us want to eat more. They are *marketed* in ways that are designed to make us eat more. That is the job of the people who sell the stuff, and it is the job of the people who make the stuff. And they do their job very well.

There were many good-tasting foods available to our ancestors. Look at fruits and berries for a moment. They have sugar and delicious flavors that evolved to entice birds and mammals to help the plant sow its seeds. But evolution is frugal. There is enough sugar and flavor to get us to eat the fruit, and no more. If the plant put more energy into the fruit than was required, the more parsimonious plants would have energy left over to grow taller and produce deeper roots. They would succeed, and would outcompete the more generous plants.

Once we learned how to breed plants, we changed that equation. We protected the plants that produced larger, sweeter fruit. We selected them and didn't plant their less generous cousins. The ancestor of corn is a tiny plant called teosinte. Its tiny fruiting body bears perhaps a dozen kernels. Early Mesoamericans selected for larger kernels, and a few centuries later we had corn on the cob. Such selective breeding has given us big juicy grapes, apples, plums, cherries, melons, and giant pumpkins weighing over a ton.

Wikimedia Commons, https://commons.wikimedia.org/wiki /File:Giant_Pumpkin_Festival.jpg

With very few exceptions, everything we eat has been modified to the benefit of humans, by humans. That includes everything on the so-called Paleo diet as well. Our early caveman ancestors did not have anything like the beef, pork, chicken, or eggs we buy in the grocery store, and nothing like the broccoli, cauliflower, squash, cabbage, lettuce, or tomatoes.

We did not stop there, however. We developed wheat from a puny ancestral grass into the profoundly productive crop we have today, with its huge stores of fast-digesting starch. We found ways to distill the juice of the fibrous stalks of another grass, the sugar cane, into the pure white sucrose crystals we buy by the 5-pound bag. We crush and press seeds and nuts to extract the pure oils. Sometimes we extract the oils with solvents to get the last tiny drops.

We have a huge industry to produce salt from brines and mines. In fact, salt was once so precious that it was used as currency for Roman soldiers, giving us the word "salary" and the expression "worth his salt."

Think of all this energy we've devoted to the production of food. But now we have more food than we can eat. Starvation still abounds, but it is not because there is not enough food. It is because we demand payment for the food, and not everyone can pay. Hunger is a problem of poverty, not of agricultural production.

The people who make the processed foods in the grocery store carry on this long tradition of invention. But now their efforts are not aimed at producing more food. There is plenty of food, and it is not very expensive. Potatoes are 28 cents a pound as I write this. So is sugar. The corn we feed to cattle is 45 cents a pound. But if we grind that corn into flour and add some sugar, we get Kellogg's Corn Pops at $4.33 per pound. People don't crave corn. Even people who crave sugar don't eat a lot of it straight out of the bag. But people will consume Corn Pops all day long if you keep a full bowl in front of them. They eat Corn Pops like the candy they are. They are pretty. They have a surprising amount of salt in them, hidden by the sugar. They have added soybean oil in just the right amount to add richness. They are light and fluffy, but crunchy. They are incredibly sweet. And they don't satisfy hunger. In fact, they make you hungrier. *By design*.

When you take something that costs 50 cents a pound and sell it for eight times that much, you want to sell as much as you can. You do that by learning what people can't stop eating.

Why can't we stop eating? There is the sugar, of course, which is damaging our satiety sensors. But there is something else.

Much of what we have been discussing so far has been about our body's natural controls over how much we eat — the various signaling molecules that make us hungry or tell us we have had enough. But we have found out how to make people eat when they aren't hungry. We have found out how to make them eat even when they are so full it hurts. After that Thanksgiving meal, when the mere sight of more mashed potatoes makes you feel a little queasy, out comes the pie à la mode, and everyone digs in.

There are two parts of the brain involved here. The homeostasis part of the brain is telling us we don't need to eat. But the reward side of the brain is anticipating that dopamine rush that happens when we eat the right combination of sugar, salt, and fat. Our processed food manufacturers have learned this combination, and more. They talk about mouthfeel. Crunch. They engineer the food to quickly dissolve on the tongue to get the sugar to the brain as quickly as possible. They study aromas that enter through the nose, but also aromas that enter the nose from the back of the mouth, where we breathe.

There was a time when it was legal to add cocaine to soft drinks. That is, of course, how Coca-Cola got its name. It was a drug delivery system. But now that company has teams of scientists studying how to make people buy and drink more sugar water. Add a bit of salt (15 milligrams per can), hidden by 38 grams of sugar per can, and people will drink more, since it no longer quenches thirst. But they don't sell only cola. Grape soda has 56 milligrams of salt. Root beer has 48. Orange soda has 45. To hide that salt, these drinks have even more sugar, with cream soda coming in at over 49 grams of sugar per can (12⅓ teaspoons).

When drinking doesn't satisfy thirst, and all the calories in the drink come from sugar, which doesn't let you know when to stop, it is easy to sell 128-ounce Team Gulp—a full gallon of sugar water. The average stomach can hold only 30 ounces, so you'll be gulping those 1,489 calories of sugar for a while. But even one stomachful of soda has 350 calories. Ten of those make a pound of fat. If you do it once a day, in a year you will have gained 36 pounds.

When most of the processed foods that are so prevalent contain so much sugar, our tongues and brains adjust, and it takes more sugar to make something taste sweet. So things keep getting sweeter. But sugar is put into foods for more than just the sweet. It thickens pastes to make them extrude from chip and candy processing machines. It adds the crunch to those Corn Pops. It binds the ingredients in crunchy granola together. It hides the taste of salt, as we saw with soda, but also with many snack foods that actually taste salty. Without the sugar, they would taste inedibly salty. Sugar masks the sour of lemons in lemonade. It hides the bitter taste of caffeine, the drug added to soda to make it more addictive than plain old cocaine.

Adding salt to the soda doesn't just prevent it from quenching thirst. Our tongues have sensors for salt, and our brain's reward system kicks in when we get that formerly rare commodity, just like it does with sugar. Having two rewarding tastes in the same package makes that product much more desirable than if there were only one or the other. Add some fat to your sugar and salt and you get caramel corn, which is also crunchy and delivers the calories extremely fast, furthering the reward.

A chocolate chip cookie is a delivery vehicle for sugar, salt, and fat. You might think you like chocolate, but plain cocoa powder is no one's favorite food. It has the flavor but without the sugar, salt, and fat. Add some fat and you get baking chocolate. Better, but still basically inedible. Add 10 percent sugar, and people will eat it. Add 20 percent sugar, and people will pay money for it. Add the right amount of sugar, salt, and fat and you get a candy bar, which is engineered to make you want to eat pounds of them when you aren't even hungry.

If there is a food that you will eat even when you are not hungry, you probably don't want to have that food close at hand when you are trying to lose weight. Throw away the cookies, ice cream, chips, sodas, etc. Just throw them away. No one else needs them either.

Sometimes it is a food you might have considered not to be a problem, or you may even have considered to be healthy. That wonderful sourdough bread. Toast and jam. (OK, you *did* know the jam was mostly sugar.) Chocolate milk. Energy bars. If you are eating it when you aren't hungry, get rid of it. It is not on your side in the weight-loss battle.

Whenever you find that you are searching for something to eat, stop and think. The mere fact that you are still searching tells you that you are not hungry. You are looking for pleasure from *food*, instead of from any of the hundreds of other things you could be doing that give you pleasure. You passed by plenty of good food in your quest, and any of it would have fixed a hunger problem. You don't have a hunger problem. You may have an addiction.

ADDICTIVE EATING

The earlier you expose children to sweet tastes, the more they will crave them later in life. Infant formula, which is more than 10 percent sugar, predisposes children to become obese adults. The more sugar a pregnant woman eats, the more it crosses through the placenta to the unborn child. This developmental, or fetal, programming affects how the child reacts to food after birth and into adulthood.

The deliciously engineered processed foods hit all the reward notes in the brain. We eat them and overeat them. Eventually, we become accustomed to the high levels of sugar, salt, and fat, and we need more to get the same reward. This is addiction.

Suggest that someone give up anything that has sugar in it. Even for a day. Most of us are addicted to sugar, and the thought of giving up sugar in our morning coffee makes us anxious. Give up all desserts? No ice cream? No chocolate? People have told me life would not be worth living. But for millions of years, it clearly was. We have only recently invented these addictive foods.

There are three signs of addiction: craving, bingeing, and withdrawal. If you crave potato chips or cookies or ice cream, that is a sign. If you have ever eaten a whole package of some food that claims to contain 4 or 5 servings, that is a sign. If you can't stop thinking about your favorite food when it isn't available for some reason, that is a sign. You are an addict.

Addicts will consume their chosen object of desire whether or not they need the calories it contains. They will get fat. They will continue to get fatter over time. Feeding an addiction just makes it worse. If you are an addict, you need to stop.

Some of us can quit cold turkey. We are lucky enough to be born with genes that make us a little less prone to addictive behavior. For most people, quitting is hard. They try to taper off, and some succeed. Others need more help—a support group, a retreat, or some other assistance. But it can be done. You can become free of your addiction to french fries or cookies or milk shakes.

It would probably be difficult for a recovering alcoholic to be a bartender. To control his addiction, he might want to stay away from alcohol. The same logic applies to a food addict. Don't bring your addictive foods home. Don't have them around the house. If you think you are depriving the rest of your family, you are letting your addiction fool you. They don't need it either, and you don't want them to become addicted. If they have to go out to get their treats, they will be less likely to become addicted. They will still love you, I promise.

If you always associate an addictive substance with a place, an object, or a person, then when you see the other thing, you want the addictive substance. The coupling is often two addictive things: alcohol and cigarettes, cigarettes and coffee, coffee and a doughnut. Other times it is just a situation: a movie and candy from the concession counter, television and a bag of chips. These are called reinforcers. Cigarette manufacturers learned long ago that constant reminders everywhere sell more cigarettes—billboards, magazines, television ads. There is always more advertising for addictive things than for any other type of product, because reminding you about your addiction makes you buy more of it. Soft drink ads are everywhere, not to get you to switch from Coke to Pepsi (that doesn't happen), but to make you think about the addictive substance.

Just reading this book is going to make you think about food. You might read all about why some foods are better than others, and which ones to stay away from, but your mind will dwell on your particular addictive food. If you find you have associated something with food, consider changing your habits. If you always eat something while watching television, replace the food with a glass of water you can sip to keep your hands and mouth busy with something nonaddictive. If you always pass that pizza restaurant on the way home, try taking a different route. Cancel your subscription to *Gourmet* magazine.

Addictive eating may be one of the main reasons why the obesity epidemic has affected the high end of the BMI range even more than the rest. We define obesity as a BMI over 30, and morbid obesity as a BMI over 40. Obesity has doubled since 1980, but morbid obesity has quadrupled in that time. As food manufacturers have learned more about how to make foods especially palatable, they have not only caused people to become addicted, but they have made it especially hard for those people who are genetically more likely to get pleasure from addictive foods (Davis et al. 2009).

There are two mechanisms that can make people prone to addiction. Some people can have a greater tendency than others to feel reward, or they can have a lower tendency. In the first case, they become addicted because they get great feelings from pleasurable things. In the second case, they compensate for the lack of reward by consuming larger quantities of pleasurable things. As we have seen in many other places in this book, there are different subtypes of obesity, and thus there will need to be different approaches to solving the problem.

METABOLIC RATE

How fast we burn calories is obviously an important aspect of any weight-loss program.

Metabolic rate is often measured as either the total energy expenditure or the resting energy expenditure. In a study of three diets (low fat, low glycemic index, and very low carbohydrate), the low-fat diet caused a decrease in *resting* energy expenditure of 205 calories per day. The low glycemic load diet caused a decrease of 166 calories per day, and the very low-carbohydrate diet caused a decrease of 138 calories per day (Ebbeling et al. 2012).

Total energy expenditure was similar—423 fewer calories burned on the low-fat diet, 297 fewer on the low glycemic index diet, and 97 fewer on the very low-carbohydrate diet.

This explains why the low-fat diet is less effective at maintaining weight loss than the very low-carbohydrate diet. We just stop burning as many calories. Four hundred twenty-three calories less per day is a lot. To make up for this, of course, we have to eat 423 fewer calories, when our body is complaining that it is already starving. (The study involved people who had just lost 10 to 15 percent of their body weight.) It is no surprise that the low-fat diets are not sustainable.

If we have just lost 15 percent of our body weight on a very low-*carbohydrate* diet, how can we stimulate our resting energy expenditure to make up for those 97 fewer calories per day?

Some things you can do to raise metabolic rate:

- Eat enough. Once you have lost the weight, stop dieting.
- Expose yourself to cold. Exercising in 60°F weather increases brown fat heat generation. Sleeping in a 66°F room also raises metabolic rate.
- Eat medium-chain triglycerides. Replacing some of the fat with MCT oil improves metabolism.
- Build more muscle mass. The muscle cells all need to burn calories just to survive.

- Eat hot peppers (capsaicin) or rub on some capsaicin pain-relieving cream.
- Drink green tea or take green tea concentrate pills. Ninety milligrams of epigallocatechin gallate (EGCG) can boost metabolism by 80 calories per day (Dulloo et al. 1999). Don't consume more than 400 mg per day.
- Drink or eat yerba mate (Y. Kang et al. 2012).
- Do interval training—walking plus 60-second sprints.
- Get 0.8 grams of protein per kilogram of body weight.
- Increase fiber intake.
- Stand instead of sitting. Good for 50 calories per day.
- Eat omega-3 fats. Reducing inflammation helps the thyroid.
- Get enough sleep, but at night, not during the day.
- Drink more water.
- Drink caffeine (but not too much).
- Reduce stress.

EXERCISE

Exercise alone does little to reduce BMI. But a low-calorie diet without exercise does not reduce fat as well as a low-calorie diet *with* exercise. Just as people have different weight-management problems, people react differently to exercise. People vary. Some lose more weight from exercise than others. But even those who lose only a couple pounds due to the increased exercise benefit from improved aerobic capacity, improved blood pressure, reduced waist size, and improved mood (King et al. 2012).

For two-thirds of the population, exercise does not make them hungry and does not cause them to eat more, even though it burns more calories than they otherwise would have burned (King et al. 2012). This was found to be true in overweight and obese subjects but not lean subjects, who did eat more in order to protect their already low stores of fat.

Exercise makes your muscles more sensitive to insulin. When the pancreas makes insulin to lower blood sugar, the muscles listen to the signal and burn more glucose. This leaves less glucose to be stored as fat.

Exercise reduces stress. Less cortisol is released into the blood. Cortisol makes you hungry. Exercise reduces appetite, because there is less cortisol in the blood. But it also improves your sensitivity to your body's appetite control mechanisms (Long, Hart, and Morgan 2002). Sedentary people lose the ability to tell when they have already eaten enough (Walleghen et al. 2007). Exercise restores that ability, so we don't overeat. The hunger hormone ghrelin needs to be activated by attaching a medium-chain fatty acid to it. If the muscles have used up all the free fatty acid, the ghrelin is not active, and you aren't hungry. There is also an increase in the hunger-reducing hormones peptide YY, glucagon-like polypeptide-1, and pancreatic polypeptide (King et al. 2012).

Exercise speeds up carbohydrate metabolism, so there is less carbohydrate in the liver to be converted into fat.

Some types of exercise build muscle. Muscle is your friend when you are dieting. Muscle tissue needs to burn calories just to survive. The more muscle you have, the more calories you burn, even while you sit at the computer or watch television, or even while you are sleeping. Creating new muscle is work, so be prepared to do some work. Feel good about it. Having a workout partner helps a lot on those days when you feel like just skipping all that work. Make it a social occasion.

Some forms of exercise are not as much work, but they help in other ways. Take a 20-minute walk. That will still reduce stress, reduce appetite, burn up liver glycogen stores, and make your muscles more sensitive to insulin. While you are walking, you aren't eating. You don't need to make it a workout that you dread. A nice walk in the park, a conversation with a walking partner, some sightseeing in a new area—all these things will make the exercise fun and hopefully something you look forward to.

You can make any errand into fun exercise. A trip to the mall can include a couple extra laps around it. (Stay away from the Cinnabon!) Park farther away. The parking is easier to find there anyway. Maybe you can bike somewhere instead of drive.

Exercise has been found to affect how people think about food. After exercise, people find it easier to restrain their appetite and retain their inhibition to eating high-energy foods (King et al. 2012). They find it easier to overcome opportunistic eating and have more control over their eating.

While the increase in lean muscle mass is the most pronounced effect of exercise on metabolic rate, there is a hormone called irisin that is stimulated by exercise, and this hormone affects fat burning (thermogenesis) in brown fat, beige fat, and to some extent white fat. It is through this mechanism that mice can be protected from age-induced obesity.

TELEVISION VIEWING

For many people, the time for watching television or using computer screens is after dark. The white pixels on the screen are made up of individual red, green, and blue pixels. Getting blue light is the signal that resets our inner clocks. We evolved in an environment where blue light meant it was daytime. There is very little blue in the light of a campfire.

When our circadian clocks are disrupted, we don't sleep well. We stay up late. In the morning, we wake either to an alarm clock or to sunlight coming through a window, complete with the light from a blue sky. We end up sleep deprived.

Sleep deprivation leads to obesity, insulin resistance, diabetes, hypertension, and all the other problems we lump into the category of metabolic syndrome.

So, while we generally sit passively in front of screens, and therefore are not getting any exercise, lack of exercise is only one component of the screen-induced obesity problem. Much of the advertising on television is about food. Just thinking about food causes insulin levels to rise in anticipation of a meal. This insulin tells fat cells to lower blood sugar by storing the energy as fat. The lower blood sugar makes us hungry.

PERSONALITY

The Three-Factor Eating Questionnaire (see page 193) rates people according to three criteria, called Disinhibition, Hunger, and Restraint.

People ranking high in Disinhibition and Hunger and low on Restraint find it more difficult to lose weight than people with the opposite tendencies (King et al. 2012). When they exercise, they tend to compensate by eating more. Those with the opposite tendencies compensate for exercise by eating less, which helps them lose more weight than others. It's as if they don't want to waste all that effort, or the exercise reminds them of their determination to lose weight.

Exercise actually helped people change their scores on the questionnaire, becoming better at restraint and eating inhibition and less hungry (King et al. 2012). This effect depended on whether they lost weight due to the exercise or not, with most improvement happening when most progress was seen.

An interesting personality effect is that overweight and obese people tend to underestimate how much they are eating (Vliet-Ostaptchouk et al. 2008). Not only does this confound studies that ask people to remember how much they ate, but it may actually contribute to obesity, since you might eat more if you think you are eating less.

In a large study of 1,988 people that spanned 50 years, several personality traits were tracked against measures of adiposity (fatness) (Sutin et al. 2011). They tested for neuroticism, extroversion, conscientiousness, impulsivity, and agreeableness. People higher in neuroticism or extroversion were more likely to have higher BMIs. People low on conscientiousness also had higher BMIs. These associations carried over for body fat and waist and hip circumference. The highest association with BMI and body fat was with the impulsiveness scores. People scoring in the top 10 percent in impulsivity weighed on average 24 pounds more than those in the lowest 10 percent.

Those with high scores on neuroticism and low scores on conscientiousness had the most weight fluctuation over the 50-year span, indicating difficulty with impulse control. Low scores on agreeableness were related to increases in BMI over time.

Self-discipline is usually required to stick to an exercise regimen or to a diet, or even to a meal schedule. This is a reflection of the conscientiousness personality trait tracked in the 50-year study. These people are more physically active, refrain from binge eating and drinking, and are less likely to have disordered eating habits.

Neuroticism is associated both with underweight and overweight subjects. Eating disorders associated with neuroticism include both overeating and undereating—bingeing, bulimia, and anorexia nervosa. There are two aspects of neuroticism that affect obesity: impulsiveness and vulnerability. The first is associated with high BMI, and the second with lower BMI (Sutin et al. 2011).

The associations with extroversion and agreeableness are less evident, and subject to cultural differences and gender, and different studies come up with opposite results, depending on the cohorts studied. The subjects who scored high on the activity aspect of extroversion tended to weigh less, while those who scored higher on the positive emotionality aspects weighed more.

Personality does not just affect the choices of when and what to eat. Neurotic and antagonistic people have greater physiological reactivity to stress and have higher cortisol levels, both of which are associated with higher BMI. They show higher levels of inflammatory cytokines such as interleukin-6, even long after the stressor has been removed (Sutin et al. 2011). Hostile individuals continue to eat even after they are satiated. This behavior adds up to weight gain over time.

Personality effects also occur in the reverse direction. Weight loss usually results in improved mood and quality of life and lowers the incidence of depression.

Last, personality traits do not guarantee that you will be overweight. They merely indicate a tendency. Personality traits lead to obesity only if they affect behavior, which is something we can control. While personality traits are fairly stable in any one individual, the *expression* of those traits can be changed and controlled. Knowing your personality allows you to take specific actions, such as hiding the ice cream when you are feeling stressed, or making grocery shopping a regular habit so that you don't eat at fast food restaurants night after night.

EATING HABITS

We have already discussed addictive eating, automatic eating, and eating for pleasure when we aren't hungry. But there are other eating habits that can contribute to obesity.

Before the obesity epidemic got going full force, we tended to eat differently than we do today. We ate three meals a day, and the meals were longer affairs. We didn't snack between meals. Our mothers told us that would "spoil your dinner" (i.e., we would not be hungry, and we wouldn't eat the food she had worked hard to prepare). Dinner was a time to eat slowly and discuss what happened during the day. Eating slowly gives your body time to generate the satiety signals that tell you when to stop eating.

These days, we wolf down a meal in 15 minutes or less. Eating too quickly allows us to overeat and regret it later when those signals catch up to us. We take pills to fix the heartburn caused by overeating.

We also eat much more between meals. The snack food industry is huge and spends enormous amounts of advertising money to convince us we need to eat at the first sign of hunger, or even to eat just in case we might otherwise become hungry. As we will see in the section on insulin, we need a period of a few hours between meals for the insulin levels to drop, or else we become insensitive to insulin and become obese and diabetic.

We have also lost all sense of portion control. Advertisements for restaurants stress the huge sizes of their burgers, burritos, or foot-long sandwiches. We get a full day's worth of calories on a single plate, and the restaurants compete in a race to put more calories on the plate. Because the calories are largely sugar and refined wheat flour, we can eat those huge portions without getting the satiety signals we need to know when to stop.

Many people reach for a sugary soft drink instead of water when they are thirsty. We don't need the calories—we are thirsty, not hungry. The sugar then makes us hungry, and the salt in the soft drinks makes us thirsty again. The makers of these drinks know exactly how to sell more sugared soda water.

The snacks we reach for tend to be low in fiber and low in moisture. An apple is satisfying because of its fiber and water content. It slows down the emptying of the stomach, and that makes the satiety last longer. A cookie has almost no fiber or moisture. Both fiber and moisture make processed foods hard to store and ship. And the snack food makers don't want to make you satisfied, because then you'd stop eating.

INFECTION

Being obese raises your risk of infection by interfering with your immune system. The body's reaction to infection is to cause inflammation, which is known to increase obesity. Obesity itself lowers immune response, but the imbalanced nutrition that led to the obese condition is also a factor, since good nutrition is important to almost every system in the body, including the immune system. Obesity is even a risk factor in influenza, increasing the probability of hospitalization and death (Morgan et al. 2010).

It is thus no surprise that obesity correlates with many infections. But what if the cause and effect also go in the opposite direction? What if there was a virus that could cause obesity?

Human adenovirus 36 is a virus associated with obesity in humans (Ponterio and Gnessi 2015). There are other adenoviruses that may also contribute to obesity.

In animals, there are several known viruses that contribute to obesity. Dogs have canine distemper virus, which causes brain damage that then causes obesity. We have seen how much of appetite control and energy balance is tightly controlled in the brain, and throwing a rock into that intricate circuitry is bound to cause trouble. Other infectious agents that cause obesity in animals are rous-associated virus (chickens) and borna disease virus (rats and chickens). These also cause brain damage, leading to obesity.

Adenoviruses are the only infective agents known to cause obesity in humans (as well as in animals). There are more than 50 types of adenovirus that can infect humans. The human adenoviruses Adv5 and Adv37 cause obesity in animals. The human adenoviruses Adv2 and Adv36 cause obesity in humans, but not in animals. Of these two, Adv36 has been more widely studied. Adv36 causes a 15 to 30 percent increase in body weight in monkeys.

In studies using rats, even killed virus cultures caused an increase of 25 percent in visceral fat mass (Ponterio and Gnessi 2015). Something in the virus affects preadipocytes, cells that go on to form fat cells. Those cells are stimulated to differentiate and proliferate, becoming fat cells. The virus also causes the concentration of lipids (fats) in the fat cells. The virus also reduces the burning of fatty acids and increases the creation of new fat, even in muscle cells.

A hallmark of this form of obesity is low blood cholesterol and low blood triglycerides, both of which are more commonly *raised* in other forms of obesity. The fat is also more subcutaneous fat than visceral (central) fat. Adv36 also improves glucose uptake, giving infected people and animals better control of glucose levels and better insulin sensitivity. These positive aspects of the infection—low blood lipids and cholesterol, improved glucose management, improved insulin sensitivity, reduced risk of nonalcoholic fatty liver disease—have led researchers to look into certain viral proteins as tools and possible treatments for diabetes and other aspects of metabolic syndrome. Insulin levels in infected animals are reduced, but whether that is caused by better insulin sensitivity or some other reason is not yet clear.

On the other hand, Adv36 most likely causes obesity by causing chronic inflammation. One of the inflammatory cytokines produced in response to Adv36 infection is MCP-1, which causes immune cells called macrophages to infiltrate fat cells, altering fat metabolism, leading to obesity, insulin resistance, and fatty liver disease, although those last two may be ameliorated by certain proteins in Adv36.

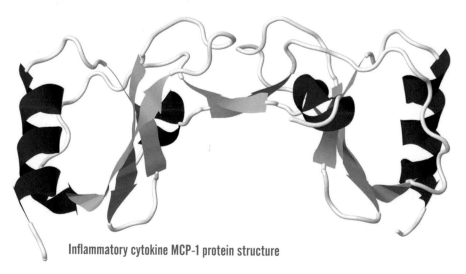

Inflammatory cytokine MCP-1 protein structure

In studies in the United States it was found that 30 percent of obese and 11 percent of nonobese adults have been infected with Adv36. It is very easily transmitted (Ponterio and Gnessi 2015), and there is a strong correlation with obesity. In US children, 28 percent of obese and 18 percent of nonobese have been infected. In Italy, the incidence of infection is 65 percent, while in Belgium and the Netherlands, the rate is only 5 percent.

Adv36 infection reduces leptin levels and norepinephrine levels. This makes the infected animals hungry, and they eat more. Since leptin is also anti-inflammatory, the lack of it further increases inflammation and thus obesity.

A killed-virus (via ultraviolet light) vaccine against Adv36 has been tested in rats (Pasarica, Loiler, and Dhurandhar 2008). Unvaccinated rats gained substantial weight after infection, while vaccinated rats were protected. There is no human vaccine at this time.

AGE

We tend to gain weight as we age. This can be the result of a continued small positive energy balance, as we get just a few more calories per day than we burn, but other factors are also at work, as there are many different interacting systems that keep our body weight stable, and over time, these can begin to fail. For example, as we age, the brown and beige fat in our bodies that burn energy to keep us warm begin to fail. Resting metabolic rate declines with age, as does the thermic effect of food (food raising body heat), and physical activity declines. Without these ways to get rid of excess energy, we store it as fat. Resting metabolic rate declines by 2 to 3 percent per decade after age 20 (Villareal et al. 2005). Most of this is due to the loss of muscle, which burns energy just to stay alive, in addition to what it burns to do work.

Average body weight and BMI gradually increase with age, reaching a peak at age 50 to 59 in both men and women (Villareal et al. 2005). After age 60, body weight *on average* declines, but this may be due to the fact that obese people don't live as long, so they aren't counted in the average as age goes up. The Framingham Heart Study saw a 6- to 7-year drop in life-span for people with a BMI over 30, compared to nonobese counterparts. The NHANES Mortality Study found that for adults in their 20s, men with a BMI over 45 could expect to lose 13 years of life, and women 8.

In addition to mortality taking its toll on these averages, the fat-free mass (bone and muscle) declines with age (up to 40 percent between ages 20 and 70), while fat mass increases continually with age. So the body weight might peak at 60, but the fat never peaks. Again, this is averaged over large numbers of subjects. It does not mean that you personally will continue to get fatter. It just means that the same thing that is causing the obesity epidemic continues to work all our lives. For most of us, we can make diet and lifestyle changes that eliminate the risk. Exercise designed to increase lean body mass (muscle) can counter this trend of losing muscle mass as we age.

As we age, the distribution of body fat also tends to change. Men tend to store more of their fat abdominally instead of under the skin, and this trend increases with age. Postmenopausal women take on male patterns of fat distribution, thus also tending toward dangerous visceral fat instead of the much less risky subcutaneous (under the skin) fat.

Two factors associated with aging, blood pressure and dementia, have been linked to insulin resistance. Curing insulin resistance can thus prevent both the high blood pressure and the cognitive decline usually associated with aging. Insulin resistance also causes fat to be stored in the muscles and liver, where it can cause serious medical concerns.

Hyperleptinemia (too much leptin in the blood, caused by leptin resistance) is common in older populations (Pérez et al. 2004). This can be fixed by dieting (caloric restriction), resulting in improved leptin sensitivity, as well as improved insulin sensitivity. Since these effects also occur in obesity, and have the same cure, we can think of obesity as a kind of premature aging.

On the other hand, too little leptin is also common as we age. Leptin is produced by fat cells and is a signal to the brain of how much fat we possess, so the brain can adjust hunger to the need for energy storage. As we age, however, the amount of leptin in the blood no longer keeps pace with the amount of fat we are storing (Moller, O'Brien, and Nair 1998). With these two effects, leptin resistance and lower leptin levels, we eat more. The fix for this is fasting, which increases leptin sensitivity and insulin sensitivity.

Aging also changes hormone levels. Human growth hormone (HGH) declines, as does testosterone.

Testosterone

We become less sensitive to thyroid hormone, hence the decline in metabolic rate. The decrease in leptin may be due to decreases in testosterone or growth hormone (Moller, O'Brien, and Nair 1998).

The age-related decline in insulin sensitivity is associated with inactivity and abdominal obesity. Physically active lean senior citizens are much less prone to type 2 diabetes as they age than those who are inactive and obese (Villareal et al. 2005).

HORMONES

There are numerous hormones and other molecules that affect energy homeostasis. This is to be expected, since anything as important as energy management will have evolved a large set of regulatory mechanisms. Some of the more important energy-regulating molecules are discussed here.

As we look at the various hormones that control metabolism, hunger, and energy, we want to remember that for every hormone there may be more than one form, and often more than one receptor, and that the hormone and the receptors are all coded for by genes. These genes can differ between individuals, so that in some people the hormones work better than in others, in ways that affect the likelihood of obesity.

Genes for hormones and their receptors are also affected by epigenetic changes to the genes, which can turn them on or off or affect how often they are expressed. Environmental changes can affect these, and they can sometimes be passed on to children.

Environmental differences can also affect how and when hormones are produced, and how well the receptors for them work. Receptors for heat and cold will affect hormones that control thermogenesis. Receptors for glucose will control hormones that affect the pancreas, causing it to produce more hormones, or sometimes fewer. Exercise causes the production of hormones that affect other body systems and usually result in yet more hormone signals to be sent.

Hormones are often antagonistic to one another, with one stimulating an effect and the other inhibiting it. There may be more than one of these co-controlling pairs, acting as backup systems and fine-tuning controls. The more important a control system is, the more likely it is that there will be several layers of control, so that if any part isn't functioning properly, other parts can rescue the organism to some extent.

This redundancy protects us from serious harm, but it also allows individuals with loss-of-function mutations in a gene to still reproduce and pass that damaged gene on to future generations.

Knowing more about your own hormone-control systems and genetics can inform diet and lifestyle choices. If you have dopamine and opioid receptors that are more sensitive than others, you may be more prone to eating addictive, rewarding foods. If your diet doesn't contain enough fiber, your gut bacteria may not be producing satiety hormones that tell you to stop eating. If you eat a lot of sweets or easily digested starches, you may be overproducing insulin, leading to a host of metabolic problems. Stress will produce hormones that have evolved to help you deal with the stressor but can cause problems if the stressor is constant. Even your gender will cause hormone differences that affect how your body stores and uses fat.

INSULIN

Many books on weight control concentrate on insulin, as if insulin alone is the cause of obesity. Books that claim that sugar or carbohydrates are the problem focus on the hormone insulin as the reason that these foods make us fat. While the truth is actually more varied and complex, insulin is indeed an important element in

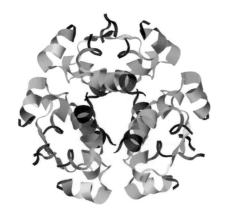

many of the problems, which are all connected in the web of how our metabolism works.

Insulin is produced in the pancreas in little clumps of cells called the islets of Langerhans, named after a medical student (Paul Langerhans) who first discovered them in 1869. Before the hormone was ever isolated, it was shown that there was an unknown substance that controlled how much glucose was produced by the liver, and that this unidentified substance came from the islet cells. The name given to the hormone reflects this. Insulin is the islet hormone.

Insulin and obesity are closely connected. Obesity is a major risk factor for type 2 diabetes through the mechanism of insulin resistance. Fat cells affect insulin sensitivity and insulin levels, and insulin strongly affects fat cells. In nonobese people, fat cells and insulin are in balance. But once we start to overeat, for whatever reason, we fill up the fat cells, and they no longer work properly. This affects insulin levels, and we get a whole range of problems, such as high blood pressure, hardening of the arteries, excess fat and cholesterol in the blood, and other conditions associated with obesity, such as polycystic ovarian disease.

Insulin acts on the ovary, causing it to produce testosterone. When a woman's liver, muscles, and fat cells become insulin resistant, the pancreas produces more insulin to compensate. This causes the ovaries to produce more testosterone, causing problems with ovulation and causing cysts to form in the ovaries. Testosterone also causes the growth of facial hair and the thinning of the hair on the scalp, effects normally found in men—beards, mustaches, and male pattern baldness. It also causes adult acne.

Fat cells (adipocytes) are very sensitive to insulin. Insulin causes cells called preadipocytes to become new adipocytes; in other words, it causes new fat cells to form. In mature fat cells, insulin stimulates glucose transport from the blood into the fat cell. It also increases the conversion of that glucose into triglycerides (fat). At the same time, insulin tells the fat cell not to break down triglycerides to produce free fatty acids, and it prevents fatty acids from leaving the cell. Furthermore, insulin tells the fat cells to absorb fatty acids and lipoproteins (proteins that carry fat, such as HDL and LDL) from the blood, and to store the fatty acids as triglycerides (Kahn and Flier 2000).

Muscle cells are strongly influenced by insulin, and they are the main tissue in which insulin stimulates glucose disposal by burning the glucose, making and storing glycogen, and by making and storing triglycerides in lipid droplets inside the cell.

Insulin tells the liver to stop producing glucose from its glycogen stores, and to produce fats and store them. It tells the muscle cells and the fat cells to increase their use of glucose from the blood.

Insulin is anti-inflammatory (Dandona, Aljada, and Bandyopadhyay 2004) and reduces harmful reactive oxidation species, but we lose this effect when we become resistant to insulin. Insulin resistance causes inflammation, which causes obesity, diabetes, and atherosclerosis.

Insulin Insensitivity

Insulin resistance has two characteristics: high levels of insulin in the blood coupled with high levels of glucose (blood sugar) in the blood. Since insulin is supposed to lower blood sugar, this state can be achieved only if cells are ignoring the insulin signal. The long-term effects are that cells run out of fuel, there are elevated levels of fats in the blood, and there is high blood pressure.

Insulin resistance predicts many diseases, such as heart attacks, stroke, ovarian cysts, macular degeneration, Alzheimer's disease, and cancer, in addition to diabetes and obesity (Monte and Wands 2008).

When we become resistant to insulin, the cells in our muscles, liver, and fat change how they respond to insulin.

Insulin resistance affects different cells in different ways. In some cases, it even affects the same cell in different ways. For example, fat cells (adipocytes) react to insulin by several mechanisms, and some are more sensitive to insulin than others are. Only low levels of insulin are needed to make fat cells stop breaking down fat. But higher levels are needed to stimulate glucose transport into the cell. Thus, even when fat cells become resistant to insulin's calls to increase glucose uptake, they maintain their fat reserves by heeding insulin's instruction to hold on to the fat (Kahn and Flier 2000).

Muscles become resistant to insulin when they get too much free fatty acid building up. This happens, for example, when alcohol is consumed. Alcohol is detoxified in the liver, where it is converted (by a complex process) into free fatty acids and very low density lipoproteins (VLDL, the worst type of cholesterol). This same thing happens when we eat fructose (or table sugar, sucrose, which is 50 percent fructose). The fructose is converted into fat, free fatty acids, and VLDL by the liver. The free fatty acids in the muscles then cause the muscles to be less sensitive to insulin. When insulin tells them to burn glucose, they ignore the signal. The glucose then ends up being stored instead of burned. The other source of free fatty acids in the muscles is excess body fat. This causes inflammation, and the fat cells release free fatty acids into the blood.

In the liver, metabolizing fructose also creates the inflammatory molecule JNK1. This interferes with the liver's receptors for insulin. The liver becomes insulin resistant. It no longer stops producing blood sugar from its stored glycogen. Blood sugar rises, prompting a rise in insulin. The insulin tells the muscles to burn glucose, but they are also insulin resistant. The high levels of insulin cause the fat cells to store more fat and to stop breaking down stored fat for energy.

Alzheimer's disease is an insulin-resistance disease (Tong et al. 2009). Long before the characteristic amyloid deposits form in the Alzheimer's-diseased brain, the cognitive impairment caused by insulin resistance is evident. Insulin resistance in the brain, leading to cognitive impairment (Blázquez et al. 2014) and Alzheimer's disease, is now often called type 3 diabetes (Monte and Wands 2008). That knowledge alone is often enough to make people give up sugar and wheat flour, as Alzheimer's disease is a death many people fear more than cancer.

Sugar is not the only thing that can cause insulin resistance. When processed meats such as bacon and hot dogs are treated with nitrites, compounds called nitrosamines are produced. Nitrosamines are also present in beer, nonfat dry milk, tobacco, and cheeses made with nitrite pickling salt. Long associated with cancer (Larsson, Bergkvist, and Wolk 2006), nitrosamines also cause insulin resistance (Tong et al. 2009). One of the drugs (streptozotocin) that researchers use to create diabetes in laboratory mice is a nitrosamine. Nitrosamine exposure exacerbates other insulin-resistance diseases, such as type 2 diabetes, fatty liver disease, and dementia (Monte et al. 2009).

Cigarette smoking, aging, and sleep problems, especially sleep apnea and night shift work, can also cause insulin resistance.

Exercise increases muscle sensitivity to insulin. Thus, to some extent, lack of exercise contributes to insulin resistance. This is separate from the number of calories expended during exercise, which is seldom anything but a small percentage of total daily calories expended.

An interesting side effect of insulin resistance is skin tags. These are small growths of skin, often in areas where the skin rubs against other skin or clothing. They are a warning sign for diabetes. People who have skin tags have higher levels of fat in the blood, higher blood sugar, higher cholesterol, and higher levels of C-reactive protein, a marker for inflammation. Insulin promotes tissue growth, and when muscle, fat, and liver cells become resistant to insulin, the pancreas produces more of it. This excess insulin has growth effects in tissues such as the skin (Tamega et al. 2010).

GLUCAGON

Like insulin, glucagon is a hormone that was named before it was actually discovered. In crude pancreatic extracts of insulin, blood sugar levels

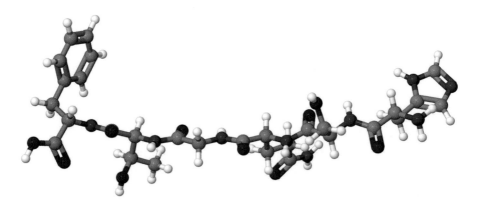

could be seen to rise before the insulin took effect and caused them to fall. This impurity in the insulin extracts seemed to act as a glucose agonist. (An agonist is something that causes an effect; its opposite is an antagonist.) The name "glucagon" comes from the words "*gluc*ose" and "*agon*ist." In the 1950s, chemists at Eli Lilly finally succeeded in isolating glucagon from other proteins in pancreatic extracts.

Glucagon is a hormone that, like insulin, is produced in the pancreas. Insulin is produced by beta cells, while glucagon is produced by alpha cells. Glucagon works in the opposite way to insulin. It raises the concentration of glucose in the blood. During fasting, when blood sugar is low, glucagon levels are highest and insulin levels are lowest. You can think of glucagon as the accelerator pedal, and insulin as the brake, for glucose synthesis in the liver.

When blood glucose levels fall, the pancreas sends out glucagon. The liver sees the glucagon, starts making glucose from its glycogen stores, and releases the glucose into the blood. When glycogen stores are low, glucagon makes the liver switch to producing glucose from breakdown products of proteins and fats.

Glucagon is an appetite suppressant, acting in the brain's hypothalamus, but also reducing levels of the appetite-stimulating hormone ghrelin, which is produced by the stomach. In obese people, however, it no longer suppresses appetite, and they remain hungry, in contrast to lean people (Arafat et al. 2013). Thus, obesity causes not only insulin resistance but glucagon resistance as well. Because glucagon still reduces ghrelin in obese people, the appetite-suppressing power of glucagon is not solely dependent on its effects on ghrelin.

There is a concept called the bihormonal hypothesis, which holds that type 2 diabetes may be as much a problem caused by too much glucagon as by not enough insulin. The two hormones act as controls on one another, and it is not always possible to say that a problem lies only with one side of the argument.

Eating protein raises glucagon levels more than eating carbohydrate does, regardless of insulin levels. This is one reason protein curbs appetite more than an equivalent number of calories of carbohydrate.

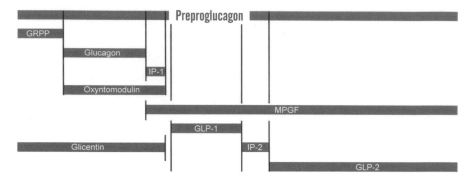

Glucagon is produced by splitting up a larger protein, called preproglucagon, into smaller pieces. Many of these smaller pieces are hormones themselves, such as GLP-1, GLP-2, MPGF, IP-1, IP-2, GRPP, glicentin, and oxyntomodulin. There are enzymes that cut the larger protein at specific points, and different tissues use different cutting enzymes to get the hormone they need from the preproglucagon molecule. Note that many of these hormones overlap one another or include parts or all of other hormones. This can allow them to bind to receptors for those other hormones. Several of these hormones have effects on energy balance and will be discussed later (pages 112 and 114).

LEPTIN

The Jackson Laboratory sits on Mount Desert Island in Maine, in the tiny town of Bar Harbor. Established in 1929, it is a biomedical research institution that studies the genetic components of health and disease. It is known for many scientific breakthroughs, but also for the more than 7,000 strains of genetically defined mice that they raise for scientific studies all over the world.

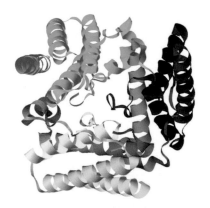

As a large breeding center for laboratory mice housed within a genetics laboratory, the lab is a perfect place for finding new mutations in genes.

The mice are examined for unique differences from their parents and other mice, and unusual variants are examined by genetics scientists.

In 1947, there was a disastrous fire on the island that burned more than half of the island over a period of 10 days. The fire destroyed most of the laboratory and most of its mice. Donations of money from around the world, as well as large donations of laboratory mice with known genetics, allowed the lab to resume its work in 1948.

Among those donated mice was one with a particular recessive mutation not seen until subsequent generations (in the summer of 1949), where deliberate inbreeding caused a mouse to be born with two copies of the gene. Such recessive genes are often genes that have become damaged and no longer code for useful proteins. When both copies of the gene are damaged, none of the protein is produced, and the effects of the lack of that protein can be seen.

In this case, the gene apparently coded for something important in the regulation of feeding and metabolism. When it was no longer there, the mouse never stopped eating and became quite obese (Ingalls, Dickie, and Snell 1950). The discoverers named the gene *ob*. Mice carrying the mutation in both sides of the DNA double helix were fat and sterile. Mice carrying the mutation in only one side were fertile and normal in weight, but a quarter of their offspring were obese, sterile mice, as you would expect from a classic Mendelian recessive gene. Mice that carry two copies of the gene are called *ob/ob* mice. If only one copy is there, the mice are called *ob/+* mice. Normal mice are called *+/+*.

In 1953, a researcher named Gordon Kennedy proposed that fat cells communicated with the brain to control feeding. In 1959, G. R. Hervey experimented with rats and found he could induce obesity by making lesions in the brain's hypothalamus (Hervey 1959). But he went further. He connected the blood system of a rat that had been operated on to make it obese with the blood system of a normal rat. The damaged rat became obese, as expected. But the normal rat stopped eating. Something in the fat mouse was producing a signal to the hypothalamus, and as the rat got fatter, more of the signal was produced. The rat with an intact hypothalamus saw the extra signal and ceased feeding. It just wasn't hungry.

In 1966, another recessive gene was discovered by Doug Coleman in mice. This one produced diabetes, and he named it *db*. Carriers are called *db/db* mice.

When the blood system of an *ob/ob* mouse is connected to a +/+ mouse (i.e., a normal, or so called wild-type mouse), the obese mouse loses weight. The signal to stop eating is coming from the normal mouse, which is not affected by the change—it continues to eat normally. When a *db/db* mouse is connected to a normal mouse, the normal mouse stops eating, but the *db/db* mouse is unaffected.

Coleman proposed that the *db/db* mice were overproducing a signal that they could not respond to, and the *ob/ob* mice were failing to produce the same signal, yet could respond to it when it came from another mouse. The *ob* gene produced the chemical signal, and the *db* gene produced the protein receptor for it. The receptor was in the hypothalamus, because lesions there caused obesity.

In 1994, after an eight-year program to find the gene, Jeffrey Friedman succeeded, and showed that the gene is expressed only in fat cells. His team manufactured the hormone and showed that it cured obesity in *ob/ob* mice, and named it leptin (Friedman 2010). For this work, Friedman won the prestigious Lasker Award and many other prizes.

Leptin is one of many hormones produced by fat cells (adipocytes). Many different tissues in the body listen for the leptin signal, so it has broad reach and several functions (Kahn and Flier 2000). It has large effects on satiety, by acting on the brain's hypothalamus. People and animals with damaged leptin function are always hungry, and if allowed to eat at will, they become severely obese. But it also has effects on energy expenditure, raising the metabolic rate so that more energy is burned (Mistry, Swick, and Romsos 1997).

Leptin makes cells more sensitive to insulin. Injections of leptin into laboratory mice lower both glucose levels and insulin levels.

Conversely, insulin also affects leptin sensitivity in the body. People who have become insulin resistant, and thus have high insulin levels, often become leptin resistant as a result (Kahn and Flier 2000). As leptin is a signal to stop eating, leptin resistance leads to overeating. Leptin-resistant people are always hungry. Since obesity causes insulin resistance, and insulin resistance causes leptin resistance, once we become obese, we are always hungry. Moreover, injecting leptin into obese patients does not help them lose weight because they are already insensitive to it.

Since leptin is made by fat cells, levels of leptin go up in proportion to the total amount of fat, not the fat percentage. Smaller obese people have leptin levels similar to larger, less obese people, when the fat weight is the same in each. Thus, the larger the body, the less likely the person is to be obese, since the leptin levels will decrease hunger. However, females have higher leptin levels than males, even after correcting for differences in body fat percentage. Male hormones appear to suppress leptin (Rosenbaum et al. 2011).

Leptin works directly on the hypothalamus in the brain to depress appetite, but it also has direct effects on other tissues and organs. Leptin is produced by fat cells, but it also affects fat cells. This is called an autocrine effect. The effect is to make fat cells less responsive to insulin, in contrast to its effects on other tissues, which is to make them more responsive to insulin (Pérez et al. 2004). When leptin levels are high, the fat cells are telling one another not to listen to insulin's signals telling them to store more fat. This natural control function breaks down when insulin levels get too high and cause leptin resistance. Leptin levels then start to rise in an attempt to compensate; this is called hyperleptinemia. This causes further insulin insensitivity in the fat cells. These positive feedback effects (vicious circles) are quite common in obesity. The good news is that since fat makes us fat, losing fat helps us lose more fat.

The effect leptin has on fat cells by making them less sensitive to insulin is not surprising, as obesity is the major factor in causing insulin resistance. Insulin sensitivity can be restored by surgical removal of visceral fat—fat inside the body, around the organs, not the fat under the skin that liposuction removes. Besides removing fat, drugs such as Metformin also work by making cells more sensitive to insulin.

In non-leptin-resistant people (people without metabolic syndrome), cells are still sensitive to leptin, and leptin injections cause cells to stop making new fat, to break up existing triglycerides, and to increase the burning of both fatty acids and glucose. It also suppresses the proliferation of new white fat cells (Harris 2014). In effect, all of this is the opposite of what insulin does.

When we eat a meal, about 40 percent of the calories go to the muscles and other peripheral tissues to be used as fuel directly. The remaining 60 percent of the calories end up in the liver, where they are stored to be released gradually between meals. If we are leptin resistant, much of the first 40 percent is not absorbed by the tissues and ends up back in the liver, where it is stored in fat droplets, creating fatty liver disease. Leptin resistance also interferes with the low-density lipoprotein receptors (LDL cholesterol) in the liver, allowing the bad cholesterol to stay in the blood, where it is prone to dangerous oxidation.

Fat is not just stored in fat cells. Almost all cells store lipids (fats) in droplets inside the cell (Susuki et al. 2011). Muscle cells store fats in lipid droplets and thus have a ready source of energy when needed. Leptin affects muscle cells by increasing the burning of fat and decreasing the conversion of fatty acids into triglycerides for storage. This is the exact opposite of what insulin does. By preventing the muscles from making triglycerides, leptin helps to keep them sensitive to insulin (Pérez et al. 2004). Leptin does not inhibit insulin-stimulated glucose metabolism in muscles (Muoio et al. 1997), so glucose burning still goes on there.

Leptin affects pancreatic beta cells, the cells that produce insulin. Like most cells, these cells store fats in tiny lipid droplets. Leptin inhibits this fat storage, both by increasing the rate at which fatty acids are burned and by inhibiting the formation of fat from free fatty acids. Pancreatic beta cells are the cells that produce insulin. They become dysfunctional when they have too much fat stored in them (lipotoxicity) (Shimabukuro et al. 1997).

Leptin affects the liver. It reduces tumor necrosis factor alpha (TNF-alpha), an inflammatory molecule produced by the liver that is associated with fatty liver disease. Leptin acts on the liver to decrease fat accumulation in the liver and to reduce the liver production of larger cholesterol particles (Huynh et al. 2012). Leptin's effects on the brain lead to downstream effects that make the liver more sensitive to insulin.

When we are leptin resistant (remember that we become leptin resistant due to high levels of insulin caused by insulin resistance, which is caused by sugar and easily digested starches), the liver has to process the calories that the insulin-resistant muscle cells refused to burn. It does this by packing the energy into LDL particles. If the meal was high in carbohydrates, the liver produces the dangerous small dense form of LDL that causes heart disease and stroke. If the meal was high in fat, the liver produces the benign large fluffy LDL particles. These particles are too large to wedge themselves into the artery linings, and instead they end up where they belong, in fat cells. Fructose in particular makes the most of the dangerous small dense LDL that damages the arteries.

Leptin is also involved in the metabolism drop that happens with weight loss from calorie restriction. Injections of leptin reverse that effect, as does fasting, which improves insulin sensitivity, and thus leptin sensitivity (Ramachandrappa and Farooqi 2011). Leptin also stimulates inflammation, and is thus one of the links between obesity and inflammatory diseases.

BROWN FAT

Leptin affects brown fat (brown adipose tissue). Brown adipose tissue is a special form of fat whose purpose is to generate body heat. It is most apparent in infants, since body heat is most easily lost when the ratio of skin surface to total body weight is high. The smaller the organism is, the higher that ratio is, since surface area grows with the square of the radius (thus square inches of surface), while body weight grows with the cube of the radius (it is measured in cubic inches). Adults have brown fat, but as a percentage of body weight it is much smaller than in infants.

Brown fat burns fat, and to some extent glucose, but does not use the energy for anything. Instead, it just lets it turn into heat. Leptin acts on brown fat to stimulate this effect, called *thermogenesis*. Burning calories for heat makes them unavailable to be stored as fat. Activating brown fat lowers blood triglycerides and cholesterol, reduces atherosclerosis, and makes cells more sensitive to insulin (Chondronikola et al. 2014). Brown fat activation increases levels of adiponectin, a hormone that regulates glucose and fatty acid burning. Eating too much (caloric excess) normally activates brown fat, as a way of getting rid of the extra calories. This is called diet-induced thermogenesis.

Besides leptin and overeating, cold exposure is another way to activate brown fat. Exercising in cold temperatures—one study used 60°F for two hours (Lichtenbelt et al. 2009)—can raise brown fat activity fifteenfold. Brown fat activity is lower in obese people than in lean people. But it is not known whether the obesity is exacerbated by low brown fat activity, or whether obese people have less need for thermogenesis due to the insulating nature of fat or the smaller surface-to-volume ratio inherent in obesity. Nonetheless, there is evidence that obese people can activate their brown fat by exercising in cool temperatures (low 60s), and that this increases the burning of glucose and fats. Brown fat is not the only mechanism involved. Muscle cells, being much more numerous than brown fat cells, are where the most thermogenesis occurs during exercise in the cold.

One study found that when activated by cold, brown fat consumed an extra 79 calories for every 15 milliliters of brown fat, for as long as the cooling was in effect (Cypess et al. 2012). As the average test subject had more than 50 milliliters of brown fat, this amounts to increasing metabolic rate by about 260 calories.

Mirabegron, a drug used to treat overactive bladder, has been shown to activate brown fat (Cypess et al. 2015). It is estimated that a 200-milligram dose could lead to an average loss of 22 pounds in three years, based on a 200 calorie per day peak energy expenditure increase.

Exercise itself may help to convert the progenitors of white fat cells into brownish fat cells (Harms and Seale 2012). These converted white fat cells are known as beige fat. This beige fat is stimulated by the hormone melatonin, the sleep hormone. Getting a good night's sleep is important for weight management. Exposure to blue light after the sun goes down interferes with circadian rhythms, preventing melatonin production. Television and computer screens are the most common source of blue light exposure in the evening.

The polyphenols in fruits like dark grapes and berries aid in converting white fat into beige fat, and increase thermogenesis (S. Wang et al. 2015).

One such polyphenol is resveratrol, found in red wine and red grapes. The best sources of polyphenols are blueberries, strawberries, raspberries, grapes, and apples.

In some people (44 percent of Europeans, 5 percent of Africans), a genetic mutation prevents the conversion of the progenitor cells of white fat from turning into beige fat. These people are more prone to higher BMI and obesity than those who do not carry the mutation.

A class of diabetes drugs called thiazolidinediones (TZDs) can activate thermogenesis in beige fat cells, as can cold exposure and the exercise-induced hormone irisin. Unlike brown fat, beige fat does not burn energy until it is activated. Without activation, the beige fat cells act like white fat cells. This ability to switch between energy storage and energy dissipation seems to be unique to beige fat cells.

Another activator of beige fat cells is the immune system. Exposure to cold induces immune cells called eosinophils to release what are called type 2 cytokines, such as interleukin-4 and interleukin-13. These molecules are produced by a class of hormones called catecholamines and are part of the "fight-or-flight" systems, generally controlled by the adrenal glands. Catecholamines include adrenaline, norepinephrine, and dopamine. Drugs that block these hormones, called beta blockers, can interfere with the activation of beige fat and contribute to obesity.

ADIPONECTIN

In 1996, only two years after the discovery that fat cells secreted the hormone leptin, researchers in the Japanese Human Genome Project were studying gene expression in adipose tissue (fat), looking for other hormones secreted exclusively by this tissue. This was the first truly systematic analysis done on gene expression in fat cells, and it hit pay dirt with the discovery of the gene apM1 (adipose

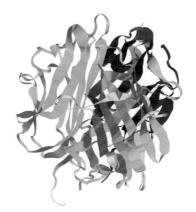

most abundant gene transcript 1), so named because it occurred most abundantly in adipose tissue. The researchers found it near a location on chromosome 3 that had been linked to diabetes susceptibility (Maeda and Matsuzawa 2014).

The protein that apM1 codes for is the hormone adiponectin. Receptors for the hormone are found in the muscles (AdipoR1) and in the liver (AdipoR2). Adiponectin is found in forms called multimers, so named because they contain multiple copies of the protein bound together. The two receptors respond to different forms of the protein, with skeletal muscle binding to a globular form, and liver cells binding to the full-length form.

One of the many hormones produced in fat cells, both brown and white, adiponectin is an important regulator of glucose and fatty acid metabolism. Low levels of adiponectin are associated with metabolic syndrome: obesity, diabetes, insulin resistance, and so on.

Adiponectin acts in the liver to slow the production of blood glucose. It acts in muscle cells to increase glucose uptake from the blood. It promotes the burning of fat and clears fat from the blood. It makes cells more sensitive to insulin, and it reduces the production of the inflammatory molecule tumor necrosis factor alpha (TNF-alpha).

When we have lots of fat on our bodies, we produce lots of leptin. Leptin reduces the levels of adiponectin. We have more fat cells, but they are producing less adiponectin. Since adiponectin is needed for insulin sensitivity, getting fat makes us become insulin resistant. Moreover, it is the harmful visceral (central) fat that lowers adiponectin levels the most, while subcutaneous fat has a much lower effect (Maeda and Matsuzawa 2014). The visceral fat inhibits adiponectin production.

When we lose fat, leptin levels fall and adiponectin levels rise. We might become hungrier, since the leptin signal is down, but we are burning more blood sugar and blood triglycerides, we are more sensitive to insulin (Yadav et al. 2013), and we have less inflammation.

Besides making cells more insulin sensitive by binding to the receptors AdipoR1 and AdipoR2, adiponectin makes us burn fat, which reduces free fatty acids in the muscles. The muscles thus become more insulin sensitive, since fatty acids interfere with the insulin signal. The insulin can then tell the muscles to burn more glucose, and to store excess glucose in the muscle cells as glycogen for use later.

Adiponectin protects arteries from atherosclerosis, and excess leptin levels damage arteries by increasing inflammation. The ratio of leptin to adiponectin in the blood is thus an indication of atherosclerosis risk, and can be tested.

The endothelial cells that line our arteries have adiponectin receptors, and they release nitric oxide when there is enough adiponectin in the blood. Nitric oxide relaxes the artery walls, allowing them to open more, thereby reducing blood pressure (Maeda and Matsuzawa 2014).

GHRELIN

The image to the right is the precursor protein for ghrelin. The aqua-colored part in the center is ghrelin.

The discovery of the hormone ghrelin was a long time coming. It started in 1962 with a program to isolate hormones that affect the hypothalamus, which include growth hormone–releasing hormone (GHRH), which controls the release of human growth hormone (HGH) (Bowers 2012). This was aided by another program that made synthetic peptides (small proteins) designed to fit into receptors for several other hormones, including opioids. Since opioids were known to stimulate the release of growth hormone, these synthetic peptides were tested to see if they could also stimulate the release of growth hormone.

Between 1976 and 1980, the program created a set of growth hormone–releasing peptides (GHRPs) that, while not naturally produced in any organism, nonetheless fit into the receptor and released growth hormone. By 1982, GHRP-6, a very effective synthetic hormone made up of six amino acids, had been created. Despite years of trying to find the natural hormone (GHRH) in hypothalamus tissue, in that same year (1982), it was finally found in a pancreatic tumor, which was releasing abnormally large amounts of it.

By 1989, it was becoming apparent that GHRP-6 and the natural hormone GHRH were actually activating two different receptors in the brain and involved different mechanisms.

In a completely surprising discovery, it was found that GHRP-6 increased food intake in rats when it was directly administered to the brain. Up until this point (1995), the work had all been related to growth hormone stimulation in the pituitary and hypothalamus. Combining GHRP-6 and the natural GHRH hormone magnified this appetite-increasing effect.

In 1999, a group of Japanese researchers, having unsuccessfully worked for years to find the natural peptide that fit the growth hormone receptor, noticed that the receptor had similarities to another receptor and concluded that whatever peptide fit the second receptor might also fit the first. The second receptor (GPR38) was expressed in the stomach, so they looked for its hormone in stomach tissues.

That is where they found the long-sought natural hormone that stimulated the release of growth hormone. They say they named it ghrelin, not because it was a **g**rowth **h**ormone–**rel**easing hormone, but from the Indo-European root "ghre," which means "grow." Other hormones, such as insulin, adiponectin, oxytocin, oxyntomodulin, and irisin, end in "in," so we get "ghrelin" (Kojima and Kangawa 2013).

The hormone ghrelin is produced in the stomach and acts in a way opposite to leptin. While leptin produced in fat cells decreases appetite, ghrelin increases appetite. Ghrelin also decreases the rate at which we burn glucose and fat. It tells the body it's running out of food, so stop wasting it.

When we become insulin resistant, the high levels of insulin in the blood interfere with ghrelin's signals. Insulin's job is to get rid of blood sugar, so it reduces the amount of ghrelin, and the muscles keep burning glucose and fat. Thyroid hormone, which regulates metabolism, has a similar effect, reducing ghrelin so that more fuel is burned. People with overactive thyroid glands burn energy faster.

As we get fatter, ghrelin production declines. This makes sense; since we have plenty of stored energy, we don't need a signal to eat more. The problem is that often the reason we get fat is that we are eating when we aren't hungry, not because ghrelin is telling us to eat. So we continue to eat when we aren't hungry, and we get fat enough to become leptin resistant, and we become hungry all the time.

Ghrelin is produced in the stomach in its inactive form. It is activated by the addition of a medium-chain fatty acid (usually octanoate) to the ghrelin molecule. During fasting, mobilization of free fatty acids occurs, and upon feeding, this mobilization is suppressed, and this leads to more activated ghrelin during fasting (Arafat et al. 2013). When ghrelin stimulates appetite, it stimulates a desire for fats more than for protein or carbohydrate (Beck 2006). Contrast this with neuropeptide Y, the powerful appetite-stimulating hormone that causes a preference for carbohydrate.

Glucagon reduces levels of ghrelin. This happens in lean and obese individuals. However, in obese individuals, this reduction in ghrelin does not cause satiety, as it does in lean people. Obese people remain hungry. They appear to be resistant to the appetite-suppressing power of glucagon, despite the lower ghrelin levels.

Ghrelin is mainly produced in the stomach but is also produced in the upper digestive tract, the lungs, the pancreas, the gonads, the adrenal cortex, the placenta, the brain, and the kidneys. Ghrelin receptors are found in other tissues, notably the lining of the arteries, where ghrelin protects against atherosclerosis (X. Xu et al. 2008).

Ghrelin is not just an appetite stimulant. It also manages the output side of the energy equation. It adjusts the amount of energy going into fat storage, glycogen storage, and into creating body heat. It also adjusts the mechanical sensors in the stomach that tell us whether the stomach is distended or not (A. Page et al. 2007). The more ghrelin there is, the less sensitive we are to a full stomach. This is why when we eat too quickly, it is so easy to overeat. The ghrelin levels have not had a chance to fall, so we are not as sensitive to the pressure signals we are getting from the stomach.

Ghrelin's actions in the brain are not limited to appetite control in the hypothalamus. Ghrelin also affects the dopamine circuits in the brain's reward center, which control addiction (Skibicka et al. 2012). Without ghrelin, there is no reward for alcohol or highly palatable foods (Egecioglu et al. 2010). Limiting desserts and alcohol to times when we have low ghrelin levels, such as when we are full, may help to reduce the addictive nature of these consumables.

Ghrelin has antidepression effects, and it decreases anxiety (Lutter et al. 2008).

Ghrelin as an appetite stimulant increases the number of meals but not the size of the meals. After a fast, eating a meal slowly enough to allow ghrelin levels to rise will thus prevent overeating.

When we gain weight, ghrelin levels drop. When we lose weight, ghrelin levels rise. This is one mechanism the body has for maintaining body fat levels (homeostasis).

Ghrelin inhibits the release of insulin after a meal. If you eat only when you are hungry, the glucose will get into your system, but the insulin will take its time. If you get carbohydrates when you aren't feeling hungry, insulin levels spike, causing all the problems associated with metabolic syndrome, obesity, and diabetes.

Ghrelin has anti-inflammatory effects and helps cells grow despite stress. It is being considered as a therapy for colitis and sepsis (Gonzalez-Rey, Chorny, and Delgado 2006). It also helps the stomach regenerate after injury to its mucosal layers.

The ghrelin receptor controls a whole raft of biological functions besides those mentioned above. It releases human growth hormone (HGH), it modulates sugar and fat metabolism, it regulates gastrointestinal motility (making food move through your digestive system better), it protects neurons and cardiovascular cells, and it regulates the immune system (Y. Yin, Li, and Zhang 2014). All these benefits come from feeling hungry. If you aren't feeling hungry several times a day, you are missing out on the benefits of ghrelin.

But wait, there's more! Ghrelin acts on the brain's hippocampus to increase learning and memory (E. Li et al. 2013). This is important to any organism that wants to remember where to find food or which food made it sick. If you want to learn something, learn it while your stomach is growling. Fasting for three or four meals makes a difference in brain function, learning, and memory. But just skipping the snacks between meals can also help, as well as lowering your insulin levels to prevent insulin resistance. Ghrelin actually helps you form new brain cells, even in adulthood. It may also be that the low ghrelin levels experienced in obese people are what lead to the non-Alzheimer's dementia that is associated with obesity (E. Li et al. 2013).

The next time you feel hungry, think of all the good things ghrelin is doing for your body, and see if you can prolong the feeling for a while. Not only will you be healthier and smarter, but you will also lose a little weight.

MELANIN-CONCENTRATING HORMONE (MCH)

In 1983, a hormone that determined pigmentation in amphibians and fish was isolated from salmon. The patterns of pigmentation were caused by the interaction between two hormones, melanin-concentrating hormone (MCH) and alpha-melanocyte-stimulating hormone (alpha-MSH). The first one (MCH) lightens the skin pigment cells, while the second (alpha-MSH) darkens them.

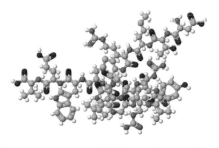

In humans, these two hormones continue their antagonistic behavior but lose their close association with pigmentation. They affect feeding (hunger), sleep, mood, and other brain functions.

MCH plays a critical role in the regulation and maintenance of sleep, especially rapid eye movement (REM) sleep. It is produced by cells in the brain, particularly parts of the hypothalamus, and consumed by cells in all parts of the brain. Malfunctions in MCH production or consumption are associated with major depression.

In addition to those functions, MCH has a role in energy homeostasis (MacNeil 2013). MCH increases appetite. Mice that have had the gene for MCH removed (MCH knockout mice) are resistant to diet-induced obesity, though due to hyperactivity and increased metabolism, not less eating. Leptin lowers levels of MCH, and injections of MCH into mouse brains raised their leptin levels by 300 percent by increasing their fat mass 100 percent.

Several different companies are working on drugs targeting MCH as antiobesity treatments. So far none have finished late phase clinical trials. Given that MCH has other important effects in the brain, such as controlling REM sleep and depression, there might not be a lot of hope for a drug anytime soon.

AGOUTI-RELATED PEPTIDE (AGRP)

This relationship between pigmentation, obesity, and metabolism is continued in another hormone, named agouti-related peptide. Agouti refers to a pigmentation effect that causes bands of color in the coats of animals by decreasing the black pigment eumelanin, thus allowing the red pigment pheomelanin to show through.

The agouti protein is responsible for banded coloring in mice and rabbits, and for the patterns in bay horses (reddish coats with black manes and tails). This change in coloration is due to interactions with the melanin system that provides pigmentation. This system is closely linked with appetite and feeding controls, as the appetite and pigment hormones are produced from the same protein, called pro-opiomelanocortin, or POMC.

The agouti protein is normally expressed only in the skin. Researchers working with obese mice found the relationship with coloration, indicating that if this gene is expressed where it normally is not, it can lead to obesity.

In 1997, two groups of researchers (Shutter et al. 1997; Ollmann et al. 1997) found a small protein (a peptide) that was very similar to the agouti protein, which was expressed eight times as much in obese mice as in normal mice. They named it agouti-related peptide (AGRP).

This hormone is expressed mostly in the adrenal gland and the brain, but small amounts are produced in the lungs, kidneys, and testicles.

AGRP increases appetite. Leptin inhibits AGRP, and ghrelin activates AGRP. Both these effects are due to the respective hormones acting in the brain on the cells that produce AGRP. Inflammatory signals also act on these cells, explaining, in part, the relationship between inflammation and obesity. In return, AGRP enhances the body's response to some inflammatory signaling molecules, in a feedback loop.

AGRP acts on melanocortin receptors in the brain, blocking the action of the appetite-depressing hormone alpha-MSH. It not only stimulates feeding behavior but lowers metabolic rate.

Stress interferes with AGRP in ways that can cause eating disorders and bingeing, and AGRP gene variants have been associated with the eating disorder anorexia nervosa, as well as obesity.

ALPHA-MELANOCYTE-STIMULATING HORMONE (ALPHA-MSH)

Alpha-MSH has several functions. It is an anti-inflammatory molecule, and it reduces fever. By itself, it is also antimicrobial, targeting both bacteria and fungi (Singh and Mukhopadhyay 2014). It protects the brain from inflammation.

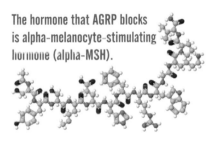

The hormone that AGRP blocks is alpha-melanocyte-stimulating hormone (alpha-MSH).

As a hormone, alpha-MSH has many functions. It regulates the melanin pigment in skin and hair, the secretion of sebum, temperature regulation, pain control, sex behaviors, and learning. In energy homeostasis, it decreases appetite. Alpha-MSH acts on the melanocortin-4 receptor, the same receptor blocked by AGRP. Leptin increases levels of alpha-MSH.

High alpha-MSH levels are associated with heart disease, obesity, sepsis, HIV, and other inflammatory diseases, but also with chronic fatigue syndrome, where it may be used as a marker for the otherwise hard-to-diagnose disease. In obese children, low levels of alpha-MSH were found, causing an increase in appetite (Vehapoğlu, Türkmen, and Terzioglu 2015).

Alpha-MSH and the related hormones beta-MSH and gamma-MSH are produced from the precursor protein POMC (pro-opiomelanocortin).

OREXINS

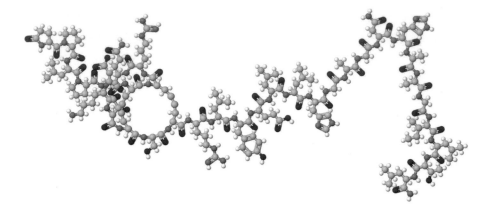

The connection between sleep, metabolism, and energy homeostasis is deep and complex. Some of the main hormones linking these physiological effects are the orexins. Discovered in 1998 by two independent research groups, both working with rats, the hormone orexin was given two names, the other being hypocretin. The hormone comes in two forms, orexin-A (shown above) and orexin-B, both small proteins (peptides), one with 33 amino acids and the other with 28.

The orexins are produced in the brain's hypothalamus, by cells that send projections throughout the brain, particularly to parts of the brain that modulate wakefulness. Orexins integrate metabolism, circadian rhythms, and sleep debt signals to control whether an animal should be asleep or awake. They interact with the dopamine, histamine, norepinephrine, and acetylcholine neurotransmitter systems in the brain.

Orexins increase appetite. They stimulate wakefulness. They regulate energy expenditure and control gut function. They are needed in order for cells to turn into the brown fat cells whose purpose is to generate body heat. Deficiencies in orexins lead to obesity by reducing thermogenesis. They also cause the sleeping disorder narcolepsy, where control of sleep is disordered, along with excessive daytime sleepiness.

Orexins' effects on energy expenditure may outweigh their effects on appetite. Orexin-deficient narcoleptic patients are obese, indicating that the lack of energy burning is more important than the lack of appetite.

Orexins promote cravings for food but also inhibit the satiety signals that usually accompany feeding, thus increasing meal size.

Leptin, the satiety hormone produced by fat cells, lowers levels of orexins. Blood glucose also reduces orexin levels. (Hypoglycemia raises them.) Ghrelin, the hunger hormone produced in the stomach, increases orexin levels.

Because sleep deprivation increases orexin levels, the body responds by increasing energy expenditure and by increasing feeding. Sleep deprivation causes a profound lack of energy as a result. Drugs (such as Merck's Belsomra) that block orexin receptors cause people to fall asleep faster and sleep longer. Similar drugs also seem to block the desire for alcohol, cocaine, and nicotine in addicted laboratory rats, thus linking orexins to cravings for drugs, sweets, and rich foods.

Orexin-A has a direct effect on fat cells, causing them to take up blood glucose and make new fat, while at the same time reducing the breakdown of fat into glycerin and fatty acids for use as fuel. It stimulates the hormone adiponectin, which turns off the production of glucose in the liver and causes cells to store glucose.

Orexins also have effects on mood, with high levels being associated with happiness and low levels being associated with sadness. Thus, a link between depression, appetite, energy expenditure, obesity, and sleep owes its effect to orexins.

CANNABINOIDS

It comes as no surprise to anyone familiar with marijuana or the marijuana subculture in popular media to know that the active component delta-9-tetrahydrocannabinol (THC, shown on the right), stimulates appe-

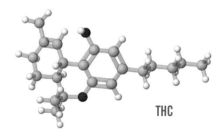

THC

tite. Users refer to this as "getting the munchies." In medicine, marijuana is used to increase body weight in patients with wasting diseases. The drug targets receptors in the brain that respond to molecules similar to THC, called endogenous cannabinoids—*endogenous* means having an internal cause or origin; that is, they are made inside the body (Marzo and Matias 2005).

Cannabinoids are under the control of leptin (Harrold and Williams 2003) and regulate energy balance, food intake, and appetite. They also act on the reward system to make eating more pleasurable. They stimulate fat accumulation and the production of fat from blood sugar (lipogenesis). Cannabinoids also stimulate production of the neurotransmitter neuropeptide Y (NPY), which is a powerful appetite stimulant, and this may be how the appetite stimulation in cannabinoids works (Gamber, Macarthur, and Westfall 2005).

COCAINE- AND AMPHETAMINE-REGULATED TRANSCRIPT (CART)

The link between appetite and addiction is clear. Not only do we use terms of drug addiction in relation to desirable foods ("addictive chocolate," "overdosing on sugar"), but we use appetite terms when talking about drug addiction ("craving cocaine"). The link is more than just a language shortcut, however. The key signaling molecules and receptors for them are shared. Addictive drugs hijack the circuits in the brain that cause us to become addicted to sugar and fat.

Cocaine- and amphetamine-regulated transcript (CART) is a small protein (peptide) that produces behaviors in the brain similar to those of cocaine and amphetamines. It also has a role in energy homeostasis and appetite. It is regulated by the appetite-controlling hormones leptin, ghrelin, and cholecystokinin. Leptin increases levels of CART, as does cholecystokinin. Ghrelin suppresses CART (Lartigue et al. 2007).

CART is an anorectic hormone (it decreases appetite). In addition to its own effects, it seems to magnify the hunger-decreasing effect of cholecystokinin. It interacts with the dopamine circuits in the brain that control desire, craving, reward, and addiction. CART stimulates the synthesis of alpha-MSH, an appetite-suppressing hormone.

CORTICOTROPIN-RELEASING HORMONE (CRH)

Corticotropin-releasing hormone is a 41-amino acid peptide (small protein) that is produced in the brain in response to stress. It is associated with major depression and Alzheimer's disease, and glucose metabolism and regulation. It stimulates the release of the stress hormone cortisol.

CRH decreases appetite, boosts attention, boosts energy expenditure, and causes anxiety. These are all normal reactions to stress, to make the animal ready for fight or flight. While the release of cortisol is anti-inflammatory, CRH itself is pro-inflammatory and may be linked to multiple sclerosis. The receptor for CRH has been linked to the euphoria produced by alcohol, and drugs that block the receptor are being investigated as possible treatments for alcoholism.

CRH has effects on energy balance beyond decreasing appetite, as it also increases thermogenesis, the burning of food to produce body heat (Richard, Huang, and Timofeeva 2000).

Leptin increases levels of CRH.

CRH increases the production of the alpha-MSH progenitor protein POMC, from which alpha- and beta-MSH, the opioids endorphins and enkephalins, and the adrenal cortex hormone ACTH, which stimulates adrenaline and norepinephrine, are formed.

ESTROGEN

Estrogen (estradiol) is the primary female sex hormone, although it is produced in both men and women and is vital to both. Estrogen affects energy balance by both accelerating 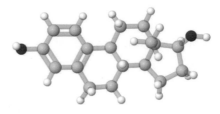 metabolism and increasing fat stores. It also suppresses binge eating, and estrogen disorders may be involved in bulimia.

Estrogen is produced primarily in the ovaries in women, but in both sexes, it is produced by fat cells throughout the body. This is the reason obese men store fat in the breast area, as women do. It is also the reason why being overweight or obese increases a woman's chances of developing breast cancer. Both estrogen and testosterone levels drop in women who lose weight (Campbell et al. 2012).

Estrogen receptors in the brain's hypothalamus exert control over body fat distribution, energy expenditure, and food intake. When these receptors are destroyed in rats, the animals eat more food, burn less energy, and gain fat mass. Estrogen is why females tend to accumulate fat under the skin, while men tend to accumulate fat abdominally, where it is associated with the harmful effects of metabolic syndrome. Postmenopausal women tend toward a more male-like fat distribution, due to changes in estrogen levels (Palmer and Clegg 2015). Girls in poor families gain more fat mass than their brothers despite similar poor diets (Grantham and Henneberg 2014), a trend not seen in wealthier families. In developed countries, both sexes are obese, but in less-developed countries, obesity is more prevalent in women. This may be due to the fact that males in developed countries are more exposed to estrogen-like substances in their environment and food supply.

Fat cells not only produce estrogen but have receptors for estrogen as well, and the sensitivity of these receptors varies with location in the body (Palmer and Clegg 2015).

Estrogen reduces the inflammation that causes obesity. Females are less prone than males to inflammatory omega-6 fats in the diet, making them more resistant to weight gain than males when fed high omega-6 fat diets (Miller et al. 2012).

Estrogen has strong effects on glucose metabolism. Both male and female mice that have the alpha receptor for estrogen knocked out have abdominal obesity, severe insulin resistance, and diabetes (Lizcano and Guzmán 2014). Estrogen also increases fat burning in brown fat tissues, increasing body heat (hot flashes are not just vasodilation).

Estrogen increases sensitivity to leptin (Lizcano and Guzmán 2014). The interplay between fat cells and estrogen is important for reproduction. Adequate fat supplies are required for fertility, and the leptin signal is how fat mass is signaled to other organs and tissues. Menopausal women are three times more likely to have metabolic syndrome and obesity as pre-menopausal women are, and estrogen/progestin-based hormone replace-ment therapy lowers visceral fat, blood sugar, and insulin (Lizcano and Guzmán 2014).

CORTISOL

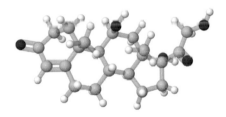

Cortisol is the stress hormone, and stress is known to be associated with obesity. So it seems natural to sus-pect that cortisol might be a cause of obesity. A disease called Cushing's syndrome, caused by taking corticosteroid medications such as predni-sone, has many of the same symptoms as metabolic syndrome, including obesity, insulin resistance, and hypertension. Chronic stress in primates has shown a fat distribution similar to that found in Cushing's syndrome (Fraser et al. 1999).

Cortisol is in a class of hormones called glucocorticoids. These are steroid hormones that regulate glucose metabolism, in addition to their immunologic effects. When we are fasting, cortisol stimulates several sys-tems that maintain normal blood sugar levels. It tells the liver to make more glucose. It mobilizes amino acids so the liver has something to convert to glucose. It inhibits the use of glucose by muscles and fat. It stimulates fat cells to break down fats, so the glycerol can be used to make glucose. Cortisol is also important in fetal development and in regulating blood pressure.

A study of a large sample of middle-aged normal subjects in Glasgow, Scotland, showed a significant association between cortisol and high BMI and high waist-to-hip ratios (Fraser et al. 1999). Cortisol levels seemed to contribute to BMI in men but not in women. In another study, levels of cortisol in the saliva of men taken right after they woke up in the morning showed a positive correlation with BMI, waist-to-hip ratio, abdominal diameter, blood glucose and insulin levels, and blood triglycerides (Wallerius et al. 2003). However, a recent study of cortisol levels in subjects with metabolic syndrome was not able to find a clear connection (Abraham et al. 2013).

Cushing's syndrome certainly causes weight gain. It also raises blood sugar levels and insulin levels and can cause diabetes. But the reverse hypothesis, that weight gain is caused by cortisol, may be true only in the case of Cushing's syndrome itself, and may not be much of a factor in everyday obesity. The association between cortisol and obesity may be the other way around, as obesity seems to cause stress. As with so many other factors relating to obesity, the cause and effect may go both ways. Stress may cause obesity, which causes more stress.

Long-term stress with elevated cortisol levels can lead to a state called hypocortisolism, where the body produces *less* cortisol than normal (Maripuu et al. 2016). The authors of that study hypothesize that long-term stress is a main contributor to metabolic syndrome and heart disease in people with bipolar disorder or recurrent depression. A link between obesity, metabolic syndrome, heart disease, and what are called recurrent affective disorders—depression, bipolar disorder—is common. People with depression are often overweight, and the link may be cortisol.

Cortisol is produced in the cortex of the adrenal glands above the kidneys. Eating and fasting both produce cortisol, as do exercise and just waking up from sleep. Stress also causes cortisol release.

Cortisol is important in the regulation and mobilization of energy reserves in the body. It selects which type of stored energy the body will use, whether carbohydrate, protein, or fat. It can move fat from one place to another. Cortisol is what tells the body to move fat from under the skin and put it deep into the abdomen. It can cause the body to make glucose from protein. It aids new fat cells to mature, and it can suppress the immune system so the body's energy stores are available to fight or flee an adversary.

The inactive form of cortisol is cortisone. The enzyme that converts cortisone to the active cortisol is produced in fat cells. The gene for this enzyme is more active in obese people than in lean people, and visceral (abdominal) fat has more of the enzyme than subcutaneous fat (fat under the skin). Visceral fat has four times the number of cortisol receptors as subcutaneous fat, and more blood flow to mobilize the cortisol. This higher sensitivity to cortisol makes the abdominal fat cells more responsive to cortisol's message to increase in size. In a study of women divided into abdominal-fat and peripheral-fat groups, those with abdominal fat were more sensitive to high carbohydrate meals (Vicennati et al. 2002).

Cortisol may directly affect feeding behavior by binding to receptors in the brain's hypothalamus, causing people to eat more foods high in sugar and fat (Epel et al. 2001). It also affects levels of leptin and neuropeptide Y, which affect appetite.

Because cortisol is associated with both higher abdominal fat and with heart disease, the waist-to-hip ratio has been found to be a better predictor of heart disease than either waist circumference or BMI.

Cortisol, like CRH, increases the production of the alpha-MSH progenitor protein POMC, from which alpha- and beta-MSH, the opioids endorphins and enkephalins, and the adrenal cortex hormone ACTH, which stimulates adrenaline and norepinephrine, are formed. All these systems interact with consequences for energy balance, appetite, and obesity.

EPINEPHRINE (ADRENALINE) AND NOREPINEPHRINE

Epinephrine is a hormone produced in the adrenal glands. It is the fight-or-flight hormone that prepares the body for dealing with life-or-death situations. It is the hormone that causes the shakiness, anxiety, sweating, fast heart rate, and raised blood pressure after a good scare.

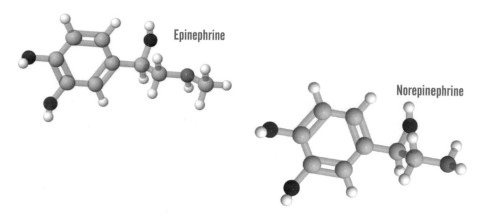

In less stressful times, it helps to manage blood sugar by making the muscles more sensitive to insulin. This is particularly useful during times of dietary excess (Ziegler et al. 2011).

Norepinephrine is a very similar molecule. It is epinephrine without the extra carbon and hydrogens (on the right side of the molecule in the illustration above). It is released by the nervous system as well as by the adrenal glands. It is also part of the fight-or-flight response, promoting vigilance, memory, and attention.

In a study of people with metabolic syndrome, levels of norepinephrine excreted in the urine were higher than normal, while levels of epinephrine were lower than normal (Z. Lee et al. 2001).

Epinephrine acts on visceral fat to release free fatty acids, which cause insulin resistance in muscle tissue (Surwit et al. 2010). Muscles are the primary way blood sugar is reduced—as insulin rises, muscles take glucose from the blood and both burn it and store it as glycogen in the muscle tissue. This is one mechanism that links obesity to diabetes.

Epinephrine moves fat from deposits under the skin into the blood. In childhood obesity, there is a decrease in the sensitivity of fat deposits to epinephrine, which may be a contributing factor, or a cause, of obesity in children studied (Bougnères et al. 1997).

Epinephrine lowers levels of the satiety hormone leptin. This causes stressed animals, and humans, to eat more (Carulli et al. 1999). Since leptin increases the burning of glucose and fat, lower levels help maintain body fat.

HUMAN GROWTH HORMONE (HGH)

Human growth hormone (HGH) is produced in the pituitary gland at the base of the brain. In children and adolescents, it stimulates growth and development, as the name implies. In adults, it increases muscle mass and decreases fat mass. Alone, it has little effect as a treatment for obesity, but in conjunction with dietary restriction, it accelerates fat loss (Kim et al. 1999).

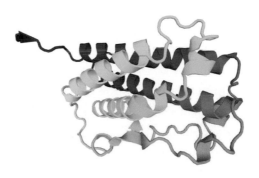

In the liver, human growth hormone is converted into insulin-like growth factor I (IGF-I). This hormone is highest during the growth spurts in puberty. As the name implies, it binds to the insulin receptor, but only about 1 percent as well as insulin binds to it. Thus it reduces blood sugar levels in the same way as insulin—just 1 percent as much.

Injections of human growth hormone alone into obese adults have not shown much effect on obesity. Weight does not change, despite small reductions in fat mass and increases in muscle mass. HGH is effective only when injected. In oral form, it is just another protein that gets quickly broken down and digested, without entering the bloodstream in any useful form. The easiest and cheapest way to increase blood levels of HGH is through fasting (Ho et al. 1988). Fasting for two days can raise HGH levels by a factor of five (Hartman et al. 1992). Fasting for 40 days in one case raised levels to 12½ times normal, without apparent side effects.

HGH counters the effects of insulin. Along with cortisol and adrenaline, it tells the body to increase the availability of glucose. Thus, raising HGH will raise blood sugar, which is safe only when fasting while awake, as blood sugar levels are low at that time. Raising HGH while not fasting can cause insulin resistance. There are also associations with enlarged hearts and increased prostate cancer incidence, which is not surprising, since it is a growth hormone and has higher levels in men.

Another way to raise HGH levels is sustained high-intensity workouts (low-intensity workouts have been found to have no effect, and resistance training was better than aerobic exercise). Vitamin D_3 increases circulating IGF-I, with similar effects—5,000 to 7,000 IUs of D_3 per week (Ameri et al. 2013).

On the other hand, abdominal fat and high insulin levels, as found with insulin resistance, reduce HGH levels (Clasey et al. 2001). So visceral fat reduces growth hormone leading to more visceral fat. In adults with growth hormone deficiency disease, there is an accumulation of visceral fat (Rasmussen 2010). Likewise, high-fat, high-sugar diets decrease growth hormone levels (Cappon et al. 2009). Drinking alcohol also lowers growth hormone levels (Prinz et al. 2009).

As mentioned earlier, sleep deprivation reduces HGH levels.

The hunger hormone ghrelin causes the release of HGH into the blood. This is why fasting raises HGH levels, as fasting leads to high ghrelin levels.

BRANCHED-CHAIN AMINO ACIDS

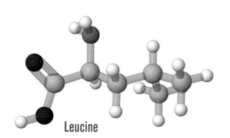

Leucine

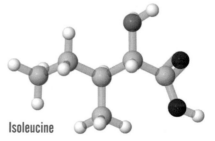

Isoleucine

Amino acids are the building blocks of protein, and there are three amino acids in particular that affect muscles and appetite hormones. These are leucine, isoleucine, and valine (as shown in the images on this page). They are called branched-chain amino acids because they are molecules that split

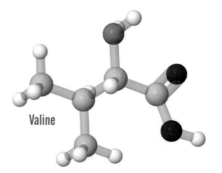

Valine

or branch in the portion that does not connect to other amino acids (the all-gray parts in the images on the previous page). Branched-chain amino acids (BCAAs) are the most abundant of the nine essential amino acids (those we cannot make, which are thus essential to the diet). Of the three, leucine is the most abundant, accounting for over 20 percent of total protein in the human diet.

These three amino acids are regulators of satiety, leptin, and glucose (She et al. 2007). They are increased in obese people, possibly due to breakdown of muscle tissue as a result of insulin resistance, and possibly because obesity interferes with enzymes for digesting BCAAs. To prevent muscle breakdown, athletes supplement with these three amino acids when doing strenuous exercise. Supplementing before and after exercise reduces exercise-induced muscle damage and promotes muscle-protein building (Negro et al. 2008). Under normal circumstances, there are enough of these amino acids in a daily diet, even a calorie restricted one, to maintain muscle mass.

BCAAs are beneficial in lean animals, including humans. They promote glucose tolerance, enhance glucose metabolism and insulin sensitivity, and decrease body fat. But in obese animals, insulin resistance cancels these effects and allows harmful accumulation of BCAAs and their breakdown products. High-fat diets can produce insulin resistance, but in combination with a high-fat diet, branched-chain amino acids produce insulin resistance even though weight gain and food intake drop (Newgard et al. 2009). They interfere with the insulin receptor molecule, making it ineffective (Chen and Yang 2015).

Each of the different BCAAs has a different effect in the body. The BCAA valine regulates fatty acid transport across blood vessel walls, allowing fatty acids into muscles, where they create insulin resistance (Jang et al. 2016). The BCAA leucine uncouples insulin signaling, also creating insulin resistance (Lynch and Adams 2014). The BCAAs leucine and isoleucine are broken down in mature fat cells and account for 30 percent of the energy used in making new fat (Green et al. 2016).

Circulating levels of these three amino acids are higher in the blood of obese people than in lean people. There are a few possible reasons for this, and one or all of them might be in effect (Chen and Yang 2015). Obese people eat more, so there is more protein to break down into amino acids. Fat cells break down BCAAs to make fat, and when fat cells are enlarged in obese people, this breakdown is impaired. Insulin resistance leads to muscle breakdown, releasing BCAAs.

Part of the damage caused by insulin resistance is due to the breakdown products of BCAAs during the abnormal metabolism of them in fat cells. These toxic metabolites of BCAAs harm the energy processing mitochondria in cells and trigger the stress response associated with insulin resistance (Chen and Yang 2015). BCAAs interact with the pancreas, changing the levels of insulin and glucagon, and may play a role in the pancreatic damage that leads to diabetes.

Higher than normal BCAA levels in the blood are one of the first indicators of insulin resistance. These levels are elevated in obese people, even those who are not yet insulin resistant. In people with metabolic syndrome, they are elevated even more than in the merely obese. Blood tests for elevated BCAAs—an indicator of problems yet to come in obese patients—may soon become a useful diagnostic tool. Catching diabetes early, before major damage is done to pancreatic cells, is important, and this diagnostic test can find problems before other symptoms are evident.

NEUROPEPTIDE Y

Discovered in 1982, neuropeptide Y (NPY) is a 36-amino acid small protein (peptide) that has many functions in the brain and nervous system. It is one of the most potent stimulators of appetite in the brain (Beck 2006), with a preference for carbohydrate consumption. It makes us eat sooner, eat more, and eat longer. It acts as a vasoconstrictor, raising blood pressure by constricting blood vessels. It causes the growth of fat tissue and blood vessels. It reduces stress and anxiety, reduces sensitivity to pain, and reduces desire for alcohol. It is one of the most abundant peptides in the brain (Beck 2006).

NPY stimulates the release of the stress hormone corticotropin-releasing hormone (CRH), and thus controls anxiety and thermogenesis (production of body heat). NPY and CRH generally have opposite effects and act as negative feedbacks to keep tight control over blood pressure and other functions. NPY raises CRH, which lowers NPY.

High-sugar, high-fat diets stimulate the release of NPY, increasing appetite and causing fat storage in the abdomen. Normally, levels of NPY increase when food is restricted and decrease upon feeding.

Beyond making us eat sooner, and eat longer, NPY motivates us to seek food, and mice will put up with electric shocks or go to more effort to get food than when NPY levels are low (Beck 2006). NPY also makes us prefer carbohydrates to protein or fat, and this effect is stronger if the carbohydrate is sugar instead of starch. Nonnutritive sweeteners such as saccharin are also preferred. It is the sweetness we seek, not the calories.

Leptin works in part by reducing neuropeptide Y (NPY) (Mistry, Swick, and Romsos 1997). Fasting for 24 to 96 hours dramatically *raises* NPY levels (Beck 2006), and feeding quickly brings the NPY levels back to normal. This is likely due to changes in blood glucose levels, which the brain can directly measure, since insulin injections, which lower blood sugar, raise NPY levels. When glucose and NPY are injected at the same time, there is less appetite stimulation.

The interaction with leptin works the other way as well. Increases in NPY cause increases in the release of leptin by fat cells (Sainsbury et al. 1996). The fat cells thus act as a brake on NPY production—a negative feedback loop.

NPY is also regulated by time. It rises at feeding times if those times are regular (Kalra and Kalra 2003).

The appetite stimulation of NPY is dependent on two opioid receptors in the brain: the mu and kappa receptors. Opioid receptors are critical to the brain's reward system and are a major part of why certain foods are addictive, but the molecules that trigger these receptors (called endogenous opioids), like endorphin and dynorphin, themselves stimulate feeding behavior, apart from their reward effects (Israel et al. 2005). NPY also interacts with other appetite control systems in the brain, such as alpha-MSH and the melanocortin system (Beck 2006) and the orexins. These interactions may link feeding to the sleep/wake cycle (Willie et al. 2001).

GLP-1

There are cells in the intestines that secrete hormones during the process of digestion that signal to the rest of the body what is coming. The cells are known by letters: the K cells secrete the hormone GIP; the L cells secrete the hormones GLP-1, PYY, oxyntomodulin, and GLP-2; the I cells secrete the hormone cholecystokinin; and the G cells secrete the hormone gastrin (Gutierrez-Aguilar and Woods 2011).

Glucagon-like polypeptide-1 is a small protein (peptide) that is broken out of the same larger protein that glucagon is cleaved from. That protein is called preproglucagon. Unlike glucagon, which raises blood sugar levels, GLP-1 causes the pancreas to secrete more insulin to lower blood sugar. Hormones that increase insulin are called incretins. GLP-1 also reduces the levels of glucagon, thus it has two ways to respond to sudden surges in glucose.

GLP-1 responds only to *rising* glucose levels, and is quickly destroyed (half-life of about two minutes) by an enzyme so it does not hang around when glucose levels are stable. It also slows acid release into the stomach, and slows the emptying of the stomach. This delays the digestion of carbohydrates, so blood sugar doesn't spike, and contributes to the sensation of satiety. In other words, GLP-1 decreases hunger.

Besides reacting to glucose, GLP-1 is also stimulated by ingested fats. When the body detects fat in the intestine, GLP-1 reduces intestinal motility, the speed with which food travels in the intestines. This allows fat more time to be digested.

The K and L cells that respond to glucose in the intestines do so using the same types of sensors found on the tongue: sweet, sour, bitter, etc. Artificial sweeteners affect these sensors as well and can cause trouble by making the body think it is getting a big meal full of glucose when it isn't. The result is a flood of insulin into the blood, which tells muscles to burn glucose, and fat cells to store it. The resulting low blood sugar makes us ravenous, and we go find some real food. And those fat cells that sucked up the glucose stay fat.

The sweet sensors on the tongue send signals to the K cells, telling them to expect glucose. This is why the satiety effects of the incretins GLP-1 and PP can occur within 20 minutes of eating, despite the fact that it takes longer than that for the food to actually reach the far ends of the intestines where the K cells are found.

Besides being secreted in the large intestine and the last third of the small intestine, GLP-1 is secreted by cells in the brain. There, it has several effects. It decreases the pleasure we get from food, and decreases the motivation to eat, along with decreasing the amount we eat and how often we eat. It slows down motor activity so we can more easily digest the meal that just raised GLP-1 levels. It can also increase anxiety and the sense of nausea.

GLP-1 increases insulin sensitivity (Sanz et al. 2015). It is required to get the high-fat diet protection provided by inulin (Vrieze et al. 2013). It also interacts with the immune system.

GLP-1 also has effects in the body that are not related to glucose metabolism. It reduces atherosclerosis by preventing the immune cells called monocytes from adhering to artery walls. It activates the immune cells called macrophages by activating the nuclear transcriptor STAT-3 (signal transducers and activator of transcription 3) (Seino and Yabe 2013). GLP-1 increases the production of nitric oxide (NO), which dilates capillaries, thus increasing blood flow into tissues. It stimulates the proliferation of the cells lining the blood vessels (endothelial cells). It reduces the production of harmful reactive oxygen species that cause oxidative damage.

GLP-1 has several cardioprotective properties. It increases glucose oxidation in heart cells, improves the amount of blood pumped, reduces hardening of the arteries, and prevents cardiac cell death. The vascular protective functions of GLP-1 are also evident in the brain, where it helps to prevent damage and neuron cell death from stroke. GLP-1 also reduces hypertension (high blood pressure), as we would expect since it stimulates production of nitric oxide, but it also helps the body to excrete excess sodium (salt increases blood pressure) (Seino and Yabe 2013).

GLP-1 reduces the amount of fat in the blood by reducing the creation of fatty acids and by limiting the absorption of fat in the intestines. In the brain GLP-1 and the related hormone GIP are important in synaptic plasticity and memory formation and have protective properties against Alzheimer's disease.

OXYNTOMODULIN (OXM)

Oxyntomodulin is another hormone cleaved from the preproglucagon protein. It is secreted along with GLP-1 and PYY when food is ingested, and it is secreted in proportion to the calories consumed. Like those other two hormones, it acts as a satiety signal, decreasing hunger. OXM reduces levels of the hunger hormone ghrelin.

Oxyntomodulin gets its name from its function of modifying (decreasing) gastric acid secretion in the stomach by glands called oxyntic glands. Oxyntic means "acid-secreting." These glands are also known as fundic glands, as they are found in an area of the stomach called the fundus.

OXM binds to both the GLP-1 receptor and the glucagon receptor, and because of that dual binding, it has greater weight-reducing power than either of those two hormones alone (Pocai 2013). It can bind to the glucagon receptor because it is pretty much glucagon plus a few more amino acids from the preproglucagon precursor molecule. By simultaneously stimulating glucose production by binding to the glucagon receptor, and halting glucose production by binding to the GLP-1 receptor, thus triggering insulin release, it makes the body more sensitive to insulin without raising blood sugar levels.

In the liver, the glucagon receptor stimulates glucose production, while in the brain (specifically the hypothalamus), the glucagon receptor starts a signal to the liver to stop producing glucose. By stimulating both, and stimulating insulin production at the same time, OXM is a potent regulator of glucose, preventing dangerous hyperglycemia.

By stimulating the glucagon receptor, OXM slows the emptying of the stomach, as glucagon does. This slows the absorption of glucose into the blood, preventing spikes in blood sugar and the resulting spikes in insulin.

In the pancreas, oxyntomodulin stimulates secretion of insulin, somatostatin, and glucagon.

Somatostatin is another hormone that reduces gastric acid secretion. In another negative feedback control function, it reduces insulin and glucagon, acting in opposition to OXM. Somatostatin also slows stomach emptying.

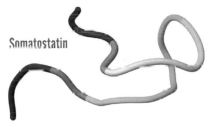

Somatostatin

When OXM stimulates insulin, it does so in proportion to the amount of glucose detected in the blood. This helps to closely control the rate of insulin production.

Subcutaneous injections of oxyntomodulin cause weight loss in both humans and rodents, increase core body temperature (burning calories faster), increase heart rate, and improve glucose metabolism (Pocai 2013). Unlike GLP-1, it does not reduce the palatability of food, but it does reduce appetite.

While surgical procedures like gastric banding and gastric bypass surgery were originally done to limit the size of the stomach, and thus reduce calorie intake, it has since been shown that one of the main reasons for the success of these procedures is that they cause exaggerated release of the hormones OXM, glucagon, PYY, and GLP-1 after meals (Chandarana and Batterham 2012). Surgery might be avoided if these hormones could be injected before meals. However, the body very quickly destroys these hormones enzymatically, so they are accurate signals of calorie intake. Making synthetic versions of these hormones that do not degrade quickly is an area of active research for drugs that control obesity and diabetes.

PEPTIDE YY

Another hormone produced by the last parts of the intestine is a 36-amino acid peptide called peptide YY. It is produced when L cells in the intestine detect feeding, and it acts as an appetite suppressant (a satiety hormone). Unlike GLP-1, it increases gut motility, speeding digestion. This gives fats less time to be absorbed, limiting the number of calories absorbed from the food. However, countering that effect, it slows down the emptying of the stomach, which allows for slower and more thorough carbohydrate absorption and prevents high blood sugar and insulin levels. It increases water absorption in the colon.

Consuming protein raises levels of PYY, reducing appetite. This is part of the reason high-protein diets help weight loss. The other reason is that high-protein diets help in thermogenesis, the use of energy to generate body heat.

People with low levels of PYY after a meal feel less satiated and tend to be obese.

Peptide YY is similar in form and function to the similar peptides neuropeptide Y (NPY) and pancreatic polypeptide (PP). Unlike NPY, which stimulates appetite, PYY and PP have the opposite effect, reducing appetite and increasing satiety.

Levels of PYY are low in the fasting state but rise quickly after a meal and stay high for an hour or two after eating. Like GLP-1, PYY levels rise long before food actually reaches the cells that produce it, indicating that sensors in the mouth or stomach are signaling for increased PYY early in the meal.

Intravenous administration of PYY has been shown to increase fat burning and increase energy expenditure, showing that these effects also occur in normal daily responses to meals. Like the other hormones that decrease appetite, PYY also works in the brain to decrease the reward from eating. Thus, it acts on both sides of the brain's obesity circuit—the homeostasis circuit and the reward circuit.

PANCREATIC POLYPEPTIDE (PP)

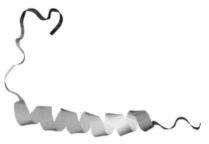

Pancreatic peptide is in the same family of hormones as peptide YY (PYY) and neuropeptide Y (NPY), sharing about half of the amino acids in those molecules. As the name implies, it is produced in the pancreas when we eat and remains at elevated levels for four to six hours afterward. The more we eat, the higher the levels of PP become.

PP decreases hunger (it is a satiety hormone). It is stimulated most by ingested protein, moderately by fat, and less so by glucose. This is one of the reasons protein and fat are more satisfying than carbohydrates.

PP reduces insulin resistance in the liver and has some small effect on insulin levels, increasing them slightly.

The appetite-controlling nature of PP was first discovered while treating children who have the genetic disease Prader-Willi syndrome. These children have insatiable hunger and quickly become obese. This hunger is relieved by infusion of pancreatic polypeptide. In some morbidly obese subjects without this syndrome, low levels of PP and low secretion of PP have been seen (Koska et al. 2004). After weight loss, these low levels of PP normalize. This means obesity causes low levels of PP, and low levels of PP do not cause obesity (Reinehr et al. 2006).

Pancreatic polypeptide is an antagonist of another appetite-decreasing hormone, cholecystokinin. Cholecystokinin stimulates pancreatic secretions, and pancreatic polypeptide turns off that stimulation. PP is stimulated by low blood sugar, eating, or just chewing something without swallowing, thus showing that the brain is the cause of the initial release of PP. As food is digested, a second phase of PP release is stimulated by the food swelling the stomach, and by the food reaching the duodenum. Levels are generally lowest in the morning and peak during the evening. As we age, levels of pancreatic polypeptide gradually rise, quadrupling between the ages of 30 and 70.

Pancreatic peptide triggers receptors in the hypothalamus and brain stem to decrease appetite. It also slows down the emptying of the stomach, reduces gall bladder contraction and gut motility, and decreases secretions from the pancreas. It increases activity and movement and increases oxygen consumption.

Besides its effects on appetite and digestion, pancreatic polypeptide reduces anxiety by slowing the release of adrenal hormones. These hormones can cause eating disorders such as bingeing and bulimia.

GIP

Glucose-dependent insulino-tropic polypeptide (GIP) is secreted in the K cells in the  large intestine and the end of the small intestine. It is another incretin (insulin stimulant), like GLP-1, but it is not as powerful a stimulant of insulin release as GLP-1. It is, however, more sensitive to glucose than GLP-1. Like that other incretin, it is rapidly degraded by a special enzyme, so it responds only to immediate rises in glucose (Gutierrez-Aguilar and Woods 2011). Like GLP-1, GIP is also stimulated by fats in the diet, but it is less sensitive to fat than GLP-1 is.

Mice lacking the receptor for GIP, or lacking the K cells that produce it, are protected from diet-induced obesity, as GIP is required for fat cells to take up fatty acids and make fat (Seino and Yabe 2013). A high-fat diet causes more K cells to grow and proliferate and produce more GIP. In turn, GIP stimulates the growth of the pancreatic beta cells that produce insulin (Trümper et al. 2009). This is increased when glucose is also present.

In people with type 2 diabetes, GIP does not stimulate the release of insulin. Interestingly, this is also true of the relatives of people with type 2 diabetes, suggesting that defective GIP signaling plays a part in that disease, and there may be genetic risk factors.

Beyond stimulating the release of insulin, GIP makes fat tissues more sensitive to insulin. In the presence of insulin, GIP causes fat cells to take up more fat (Seino and Yabe 2013). It also increases glucagon secretion when fasting or when blood sugar is low. (Glucagon raises blood sugar levels.) Thus, it stimulates insulin to fix high blood sugar, and glucagon to fix low blood sugar (Christensen et al. 2011).

The downside of GIP is not just related to its critical role in the accumulation of fat. It also enhances glucocorticoid hormones that promote inflammation, and these in turn contribute to obesity. This is slightly balanced by GIP's actions in decreasing appetite and food intake (Seino and Yabe 2013). Unlike the similar hormone GLP-1, which reduces gastric emptying, GIP facilitates gastric emptying, allowing ingested carbohydrates to spike insulin levels. This complex interaction of GIP with glucose and fat metabolism and fat cell regulation makes teasing out its total effect difficult. For example, GIP protects against diet-induced obesity and fatty liver disease.

Along with GLP-1, GIP protects against Alzheimer's disease and helps synaptic plasticity and memory formation. It is also protective against bone fractures, by interacting with cells called osteoclasts, which mine the bones for calcium (Seino and Yabe 2013). It stimulates bone formation by reducing the death of the bone-building cells called osteoblasts.

CHOLECYSTOKININ (CCK)

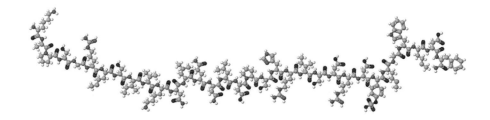

Cholecystokinin is another hormone that decreases appetite (Little, Horowitz, and Feinle-Bisset 2005). It is responsible for the digestion of fat and protein. It causes the release of digestive enzymes and bile from the pancreas and the gall bladder. It is released rapidly after a meal in response to fatty acids and some amino acids in the first part of the small intestine. It is degraded quickly, so it can act as an accurate measure of ingested fat and protein.

CCK slows the emptying of the stomach and decreases the amount of acid released by the stomach. CCK is inhibited by pancreatic polypeptide (PP), another satiety hormone triggered by protein ingestion.

Cholecystokinin stimulates the vagus nerve, which, among many other functions, controls peristalsis, the movement of food through the intestines. The stimulation of the vagus nerve can be inhibited by capsaicin, the molecule that makes hot peppers burn, and this is a key to how capsaicin helps in weight management.

Cholecystokinin is produced in the first section of the small intestine, the duodenum, and the section right after that, the jejunum. But it is also widely distributed in the brain's hypothalamus, and it is the most abundant neuropeptide in the entire central nervous system. There are two receptors for cholecystokinin, one in the gastrointestinal tract, known as CCK_A, and one in the brain, called CCK_B. CCK_A seems to be the most important receptor when it comes to regulating food intake. Drugs that prevent this receptor from working (CCK_A antagonists) increase both hunger and meal size in humans (Perry and Wang 2012).

OXYTOCIN

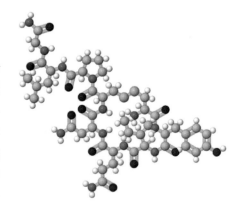

The hormone oxytocin is usually thought of in the context of social bonding. Positive social interactions cause oxytocin release from the pituitary gland, producing feelings of contentment, calmness and security, and reduced anxiety.

Stimulation of the nipples during breast-feeding causes secretion of oxytocin, which causes milk to flow and helps to promote bonding with the baby. Oxytocin is also a powerful stimulant of uterine contractions during childbirth. It has anti-inflammatory effects and promotes wound healing. It also has cardioprotective effects—it prevents heart disease (Jankowski, Broderick, and Gutkowska 2016). In obese subjects, and subjects with diabetes, it has been found that oxytocin levels were lower than in normal subjects (Qian et al. 2014).

But oxytocin has many other effects in the body, and some of them have to do with energy balance, insulin signaling, and insulin resistance. Nasal spray administration of oxytocin has been found to be effective in increasing insulin sensitivity, increasing insulin production, reducing blood triglycerides and cholesterol, and reducing fat mass (H. Zhang et al. 2013). Patients with BMIs over 28 lost an average of 10 pounds in four weeks, and over 19 pounds after 8 weeks, with no adverse side effects (24 units of oxytocin nasal spray 4 times a day, 20 minutes before each meal, and before sleeping). LDL cholesterol dropped, and HDL cholesterol rose—exactly what you want to see. One side effect of oxytocin is decreased food intake. Thus, the weight loss may not be due solely to improved blood glucose and insulin management.

Oxytocin is currently in clinical trials for diseases including autism, anxiety disorder, depression, drug abuse, and schizophrenia. The links between metabolic diseases and depression, bipolar disorder, and schizophrenia make treatment with oxytocin for obesity and diabetes an interesting research subject.

In studies of diet-induced obesity in rats, administering oxytocin reduced body weight gain by 50 percent, without changing food intake and without any change in meal patterns. This means the hormone caused a decrease in food efficiency (Deblon et al. 2011). Oxytocin also increased levels of glycerol in the blood and decreased triglycerides (fat) in the blood. It did this by increasing levels of enzymes that break fat down into glycerol and free fatty acids, and by increasing the enzymes involved in fatty acid transport. The levels of free fatty acids in the blood did not rise, indicating they were burned or incorporated into cell membranes.

Increasing the dose of oxytocin did reduce the number of meals, but not the size of the meals or their duration. The hormone thus increased satiety. The rats also burned fat in preference to carbohydrates, compared to rats given a placebo.

Oxytocin induces its own synthesis and release. When oxytocin is injected into rat brains, the rats produce more oxytocin, releasing it into the bloodstream.

Fat cells have receptors for oxytocin, and when those are triggered, the fat cells release enzymes that break down fats, and burn the free fatty acids that result. The body weight reduction caused by oxytocin is not due to its anorexic effects (i.e., not from reduced food intake) but by the breakdown of fats in fat cells (Deblon et al. 2011). In the testing on rats, oxytocin increased insulin sensitivity and glucose tolerance, independent of its effects on fat mass.

ACETATE, PROPIONATE, BUTYRATE

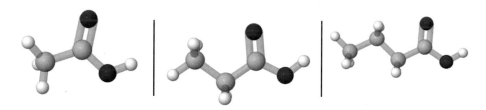

The fatty acids butyric acid and acetic acid have been shown to be protective against diet-induced obesity and insulin resistance (Lin et al. 2012), even when not reducing food intake. Propionic acid, on the other hand, has been shown to reduce food intake. Butyrate and propionate, but not acetate, induce gut hormones and reduce food intake. (When fatty acids combine with other molecules, they lose a proton and are no longer acids, and we refer to them by the name of the remaining part of the molecule.)

All three of these short-chain fatty acids are produced by microbes in the human gut, mostly the large intestine, and these microbes are strongly affected by diet.

Butyrate (the four-carbon fatty acid) is the main food source for the cells lining the intestine (enterocytes). Getting enough fiber in the diet helps to strengthen the gut barrier. This protects us from bacteria and inflammatory molecules that can otherwise leak through the intestinal lining. But butyrate also acts as a hormone, reducing inflammation, oxidative damage, and transcription factors that activate inflammatory cytokines. Butyrate gives a feeling of satiety (fullness) after a meal. Butyrate also promotes the formation of regulatory T-cells, immune cells that regulate immune function.

Mice fed a high-sugar, high-fat diet become obese and insulin resistant. Adding butyrate to the diet prevents and reverses insulin resistance.

Butyrate increases energy expenditure by modulating the mitochondria, the organelles inside cells that are responsible for managing energy.

Acetate and propionate stimulate production of peptide YY, which speeds up the passage of food through the intestines, thereby reducing the time it has to be absorbed. This reduces the amount of energy the body can harvest from a meal. Acetate and propionate also induce secretion of GLP-1, the hormone that increases insulin sensitivity and satiety.

THYROID

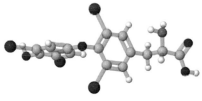

Thyroxine

The thyroid gland regulates metabolism and has interactions with insulin resistance, adiponectin, ghrelin, and leptin. Diet restriction (low-calorie diet) reduces thyroid hormone levels, reducing metabolism. Inflammation also reduces thyroid levels.

The thyroid gland produces three hormones, called triiodothyronine (T_3), thyroxine (T_4) (shown above), and calcitonin. Calcitonin regulates the levels of calcium in the body. T_3 and T_4 regulate metabolism, and thus we would expect them to have an influence on energy balance and body composition. They influence appetite, gut motility, and nutrient absorption.

T_4 has four iodine atoms, and T_3 has three, hence the names. Most of the T_3 in the body is made from T_4 by an enzyme that removes one of the iodine atoms. The hypothalamus releases pulses of a hormone called thyrotropin releasing hormone (TRH), which stimulates the pituitary gland to produce thyroid stimulating hormone (TSH), which then stimulates the thyroid to produce T_3 and T_4.

Too much thyroid hormone produces a state called hyperthyroidism, which is characterized by weight loss and hunger, along with many other symptoms such as anxiety and nervousness. Thyroid hormone affects tissues throughout the body and has many effects on homeostasis, such as the paradoxical state where the antagonists ghrelin and leptin are both reduced at the same time (Potenza, Via, and Yanagisawa 2009). When patients with hyperthyroid are treated for it, they often experience weight gain.

Too little thyroid hormone causes the opposite condition, hypothyroidism. This state is characterized by weight gain, along with tiredness and slow heart rate, among several other symptoms. Hypothyroidism can be caused by iodine deficiency (rare, now that most salt is iodized), or by autoimmune disorders, inflammation, and several other problems. It can be treated by adjusting the diet or by taking oral thyroxine pills (synthetic thyroid hormone).

If we can cure obesity by taking a thyroid pill, why aren't doctors handing them out like candy? Because hypothyroidism is only one small part of the obesity epidemic, and thyroid hormone does not cure obesity in people with normal thyroid levels. And, of course, there are side effects.

However, while an underperforming thyroid can sometimes cause weight gain, obesity itself can cause thyroid problems. Fat tissue makes many hormones, including some that promote inflammation, and it also harbors immune cells called macrophages. When fat cells are overproducing these inflammatory molecules and immune cells, the result can be autoimmune diseases, where the body starts attacking itself. Autoimmune attacks on the thyroid gland cause hypothyroidism. This in turn exacerbates the obesity problem—another vicious circle.

With the complex set of hormones and enzymes that control the secretion of thyroid hormone, we would expect to find some natural variation between people in how well the system works. Small genetic differences should affect the amount of hormone or enzyme available, and slightly alter how much thyroid hormone is available, and how well the receptors for the hormone work. For the most part, these differences are smoothed out by negative feedback control systems, such as producing more hormone if the receptors aren't always reacting to it enough. Still, we should see some variation. And a little bit of hypothyroidism, perhaps low but still in the normal range, could add up over the years into a weight problem (Biondi 2010).

Obese people produce more leptin than lean people, and leptin stimulates TRH, which stimulates TSH, thus increasing thyroid hormone levels. TSH, in turn, stimulates leptin secretion from fat cells, in a feedback loop. Leptin also raises levels of the enzyme that removes iodine atoms from T_4. Thus, leptin and thyroid hormones control one another. Leptin also increases susceptibility to thyroid autoimmunity by regulating immune processes (Biondi 2010).

In obese people, levels of TSH are at or near the upper limits of the normal range, and the higher the BMI, the higher the TSH levels. This may be because of increased leptin, but the TSH receptors in fat cells also are less expressed when the fat cells are overinflated, and so more hormone may be secreted to make up for the reduced sensitivity. This last condition is reversed by weight loss, resulting in better thyroid function. With weight loss comes a reduction in leptin, and in TSH, which causes the thyroid to produce less hormone. Because this lowers metabolic rate and fat burning, this can make further weight loss more difficult (Biondi 2010).

DOPAMINE

While the reward system in the brain is very complex, and details are still being worked out, the role of the neurotransmitter dopamine has come to be seen as what makes people *want* something. It is not the reward itself,

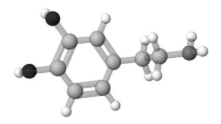

or the rewarding feeling, but rather the desire for reward that dopamine seems to create.

Dopamine is often studied in relation to addiction. People with low levels of dopamine may become addicted as a way of "self-medicating" to bring their dopamine levels up (Davis et al. 2009). This same mechanism may be involved in obesity, as there are many indications that some forms of obesity are addictive behaviors involving food and food reward.

Dopamine evolved as a way of signaling to the animal that something was important. When we eat something that is highly rewarding (see the section on opiates on page 127), the brain produces dopamine to make us remember that stimulus. Dopamine makes us *want* something.

At the same time, there is some evidence that high levels of dopamine make people more sensitive to reward, and thus more prone to eating for pleasure or even to binge eating.

Cocaine and amphetamines work on the dopamine system by increasing dopamine levels. These are highly addictive substances. Highly palatable foods full of sugar, fat, and salt are only a fraction less effective at stimulating dopamine as cocaine. It is easy to see how we can become addicted to these processed foods. When people refer to their favorite cookie as their particular form of crack cocaine, the joke is not all that far from what is actually happening.

IRISIN

Irisin is a small protein (made of 112 amino acids) secreted by the muscle cells during exercise (Bostrom et al. 2012). It increases beige fat production, and thus thermogenesis (burning fat and glucose to generate body heat). Glucose tolerance is also improved, and fasting insulin levels are reduced. Irisin is powerful: very tiny amounts have been found to increase fat burning by a factor of 50, and to protect mice from obesity induced by a high-fat diet.

The genes for irisin are highly conserved, remaining 100 percent the same in both mice and humans. Compare this to the 85 percent similarity for insulin, the 90 percent similarity for glucagon, and the 83 percent similarity for leptin. This implies that the protein activates a cell surface receptor, since very small changes in the activating protein for such receptors result in a failure to activate them. It also speaks to the importance of the molecule in survival and reproduction.

Why should exercise stimulate beige fat thermogenesis? One theory is that since shivering is one way the body protects against hypothermia, shivering muscle cells get assistance in generating body heat by activating the other mechanism of thermogenesis.

Unfortunately, polypeptides (small proteins) like irisin are easily digested when eaten, so the protein would have to be injected in order to be of any benefit. In the meantime, we can produce our own irisin by exercising (or shivering).

The statin drug simvastatin (Zocor) has been shown to increase irisin levels in the blood. Unfortunately, this may be a side effect of one of its side effects: muscle breakdown. The drug also increases liver problems and blood sugar levels, so it is unlikely to be a popular weight-loss solution.

Besides its effects on energy balance, irisin also seems to explain the effects of exercise in improving brain function and preventing depression, epilepsy, stroke, Alzheimer's, and Parkinson's disease (Wrann 2015).

White fat cells and bone cells differentiate from the same precursor cells. Irisin prevents these cells from becoming fat cells, and instead promotes the formation of bone cells. This explains in part why exercise is helpful in bone growth and maintenance and protects against osteoporosis (Y. Zhang et al. 2016).

Irisin also reduces blood cholesterol formed in the liver (Tang et al. 2016). This explains in part the role of exercise in lowering cholesterol levels.

ENDOGENOUS OPIOIDS

Endogenous opioids are your own brain's source of heroin and morphine. Those drugs act on receptors in the brain that are, in the absence of those drugs, the targets of natural molecules that are similar to the drugs. These

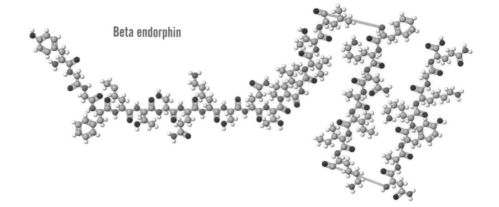

Beta endorphin

endogenous (originating from within the organism) opioids (molecules similar to opium) are the molecules that control what we *like*. This is why they are addicting. They work on the part of the brain's reward system that makes us *like* something. You may know these molecules as endorphins. (There are four types of endorphins, named alpha, beta, gamma, and sigma. The strongest of these is beta endorphin.)

When we give opioids to animals that are not hungry, they eat. The effect is selective for highly palatable foods. The activation of the brain's opioid receptors increases the reward we get from sweet and fatty foods (Davis et al. 2009).

Animals naturally choose variety in what they eat. This ensures a balance of nutrients. When we are given one type of food, we tire of it after a while. But foods made of sugar, fat, and salt trigger such reward in the brain that we don't get tired of them. We keep trying to get that opiate reward. If we block the action of opiates in the brain (using drugs developed to treat opiate addiction), the desire for this high-sugar, high-fat food goes away, since they no longer provide the reward.

It comes as no surprise that the opioid receptors are the target of much study in the fields of alcoholism and drug addiction.

Together, dopamine and opioids form the powerful reward system that processed foods and drugs like cocaine, amphetamines, heroin, morphine, and alcohol are all produced to exploit for profit. Hijacking the reward system is a very effective way to sell products.

Besides their reward functions, endogenous opioids also stimulate appetite (Israel et al. 2005).

HYPOTHALAMUS AND HORMONES

In the brain's hypothalamus, several of the hormones act on the appetite regulation systems in the parts of the hypothalamus called the arcuate nucleus and the paraventricular nucleus. Leptin, insulin, and ghrelin all have receptors in the arcuate nucleus. Leptin and insulin stimulate production of the precursor protein POMC, from which alpha- and beta-MSH peptides are formed. The MSH hormones activate the MC4R melanocortin-4 receptor in the paraventricular nucleus, which controls appetite (Pritchard, Turnbull, and White 2002).

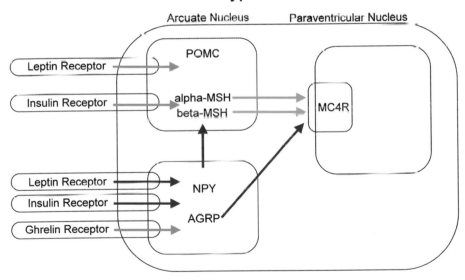

The arcuate nucleus also has receptors for leptin, insulin, and ghrelin that produce the antagonists to the MSH hormones. These are NPY and AGRP. While MSH decreases appetite, NPY and AGRP increase appetite both by acting on the MC4R receptor itself and by reducing the expression of MSH elsewhere in the arcuate nucleus. These competing systems maintain a tightly controlled balance for energy homeostasis, and subtle changes in the genes for any of these signaling molecules or their receptors can influence the risk of obesity (Pritchard, Turnbull, and White 2002).

Other influences on POMC production are CRH, CART, cortisol, and some of the interleukin-6 family of cytokine messengers. Appetite is a strongly controlled behavior modifier, and the control system is very complex, with many interacting systems providing input from many different internal and environmental signals. This means that while there are many ways for the system to go wrong, there are redundancies that minimize the effects of any particular locus of damage. Evolution has had a long time to work on this system.

POMC is broken up into a whole raft of smaller peptide hormones by several enzymes with the help of other proteins to chaperone and ferry the parts to where they belong. Thyroid hormone also affects how POMC is broken down. Any of these enzymes and proteins can have variations that make them less able to perform their function. The result is an increased risk of obesity but with very small effects. It is the combination of many of these problems that raises the risk of obesity to levels that express as increased BMI. Most of these variations have been found to relate specifically to diet-induced obesity, so an environment of highly palatable high-energy foods interacts with genetics to produce a medically important result (obesity, diabetes, stoke, dementia, metabolic syndrome, etc.).

Many of the products produced by breaking up the POMC protein are capable of interacting with the MC4R receptor. This means that if something goes wrong with the production of alpha-MSH, beta-MSH and DA-alpha-MSH (two other breakdown products of POMC) can step in. And due to negative feedbacks in the control system, if there is a problem making some hormone, the body detects that the desired effect is not happening at sufficient levels, and it requests more, thus making up for the deficiency. The levels of AGRP can also be adjusted to get the desired result, since it has the opposite effect as alpha-MSH. So while many things can go wrong, there are safeguards in place to minimize the damage.

NITROSAMINES

Nitrosamines are small molecules formed when a protein reacts with nitrites or nitrates, either through heat, as in cigarette smoking and frying, or by acids, as when nitrate-containing processed meats are digested in the stomach. Tobacco processing to create chewing tobacco

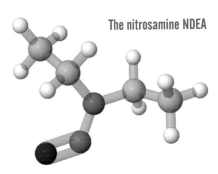

The nitrosamine NDEA

and snuff also creates nitrosamines. Beer, processed meats, and cheeses processed with nitrites are all sources of nitrosamines. Nitrosamines are also found in cosmetics and pesticides.

Ascorbic acid (vitamin C) added to foods helps to prevent the formation of nitrosamines.

Most nitrosamines are carcinogenic. They damage DNA.

In one study, early exposure to nitrosamines in rats led to obesity, dementia, diabetes, and metabolic syndrome (Tong, Longato, and Monte 2010) when the rats were fed a high fat diet. In another study (Tong et al. 2009), Alzheimer's disease, complete with its characteristic beta-amyloid plaques, insulin resistance, fatty liver disease, and diabetes were induced in rats with a single dose of the nitrosamine NDEA (shown above). Since nitrosamines are very common in processed foods, chronic exposure at a young age may well predispose humans to metabolic syndrome, diabetes, and Alzheimer's disease.

SUMMARY

The diagram shown on the next page illustrates the complexity of the system that regulates appetite, feeding, energy, and homeostasis. Proteins and peptides are shown in rectangular boxes. Those that increase appetite are shown in yellow, and those that decrease appetite are shown in blue. Arrows connecting boxes show whether a substance increases (green) or decreases (red) the effect of the target substance, cell, or receptor.

Inputs to the system, such as glucose, fructose, alcohol, vitamins, exercise, stress, fiber, oxytocin, nitrosamines, polyphenols, fatty acids, cold, and sleep, are things we have control over. We also have control over our gut microbiome, which produces some of the inputs. Genetic effects, which we generally have less control over at this time, are responsible for many of the levels of these hormones and the effectiveness of their receptors. Knowing our genetic contributions to the system can help to guide any of the interventions we make to control the system.

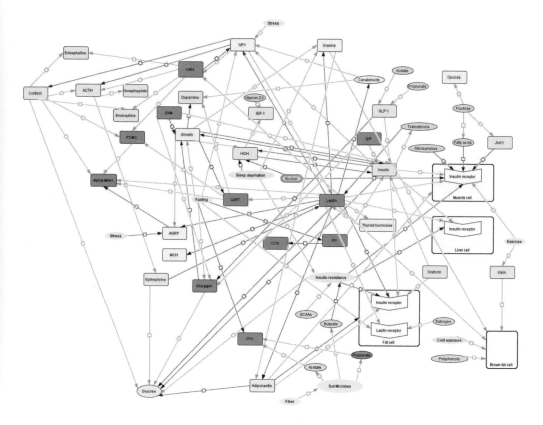

GUT MICROBES

Mice are used a lot in science laboratories because, being mammals like us, most of their biological functions are similar or identical to our own. Their other great advantage is their fast breeding rates and inexpensive maintenance costs. In many cases, It is useful to breed the mice in germ-free environments, to reduce variability and eliminate external sources of genetic material and other molecules. One thing that stands out in germ-free mice is that they are less prone to diet-induced obesity than normal mice. Germ-free mice given the microbes from normal mice started to gain considerable fat within 10 to 14 days despite the fact that they ate less food (Turnbaugh et al. 2006).

Since the mice were nearly identical genetically, and the only difference was that some were raised in a germ-free environment, scientists began to look at the microbes in the guts of the animals to see if there was a link to obesity. The gut microbes were able to break down (ferment) dietary fiber that the mice could not otherwise digest. The resulting simple sugars and short-chain fatty acids were then absorbed by the intestines. Those were then processed in the liver to make fat. Moreover, it appeared that the microbes were producing signaling molecules that made the mouse fat cells absorb more fats (Turnbaugh et al. 2006).

With the discovery of *ob/ob* mice, scientists had another model for obesity they could experiment with. These genetically obese mice could not make leptin and ate constantly. But it wasn't just that they ate more food; they also seemed to harvest more energy from that food than otherwise identical lean mice did.

When scientists analyzed the difference between the gut microbes in obese mice and those in lean mice, it was clear that the obese mice had differ-

Genetically obese mouse. *Wikimedia Commons, https://commons.wikimedia.org/wiki/File:Bioengineered_obese_mouse,_Aberdeen,_Scotland,_1998_Wellcome_L0060083.jpg*

ent gut microbes from the lean mice (Ley et al. 2005). Gut bacteria are dominated by two phyla, called the Bacteroidetes and Firmicutes.

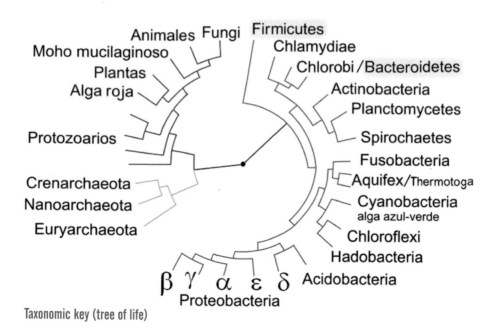

Taxonomic key (tree of life)

The obese mice had half as many Bacteroidetes as the lean mice, and an increase of the same amount in Firmicutes. Analyzing human gut microbes showed a similar increase in Firmicutes at the expense of Bacteroidetes. As humans lose weight, this shift reverts back to a more normal distribution.

At the time of this study, it was not known whether this difference helped to make the mice obese, or if it was a result of obesity. But the researchers noted that feeding antibiotics to cattle increased their growth rate, a common practice in the beef industry. Perhaps that growth was spurred by gut microbes that helped to digest fiber and convert it into short-chain fatty acids, to be absorbed into the cattle or mouse bloodstream.

Another study soon determined that this was indeed the case (Turnbaugh et al. 2006). When scientists transferred gut microbes from obese mice and lean mice into germ-free mice, the mice with the microbes from the obese mice gained significantly more weight than those that got the microbes from the lean mice. This was despite the fact that all the mice ate the same amount of food. The obese mice were getting more calories out of the same amount of food. The amount came to an extra 2 percent more calories. In humans, that amount would cause us to gain a pound every three months.

That microbes in the human gut can ferment fiber to make sugars and fats that the intestines can absorb should not be surprising. After all, we know that in cows and other ruminants, microbes break down fiber that the cow cannot digest, and the cows get a substantial portion of their calories from this fermentation. Humans lack the special stomach (rumen) where this happens in the cow, yet we still harbor a substantial colony of microbes in the lower intestine. In fact, our bodies have as many microbial cells as human cells (Sender, Fuchs, and Milo 2016). Those bacteria produce vitamins that we cannot (notably vitamin K, vitamin B12, folic acid, biotin, and pantothenic acid), as well as other important molecules. We are a symbiotic organism, host to 200 grams of microbes, mostly in the gut.

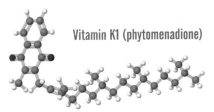

Vitamin K1 (phytomenadione)

Giving germ-free mice gut microbes from normal mice causes insulin resistance and a 60 percent increase in body fat (Bäckhed et al. 2004). Conventionally raised (normal) mice have 42 percent more total body fat than germ-free mice do, despite the fact that the normal mice eat 29 percent less food than the germ-free mice.

Conventionally raised mice have higher metabolic rates than germ-free mice (Bäckhed et al. 2004). The metabolic rates of germ-free mice were found to be 27 percent lower than those of conventionally raised mice. So the fat was being added despite less eating and more calorie expenditure.

There is a key regulator of fatty acid release from lipoproteins in fat, muscle, and heart cells called LPL (lipoprotein lipase). When LPL levels rise, cells take up fatty acids, and fat cells grow larger.

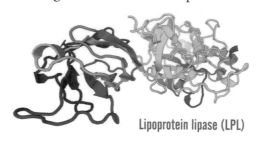

Lipoprotein lipase (LPL)

In analyzing germ-free and normal mice, it was found that there is a protein called fasting-induced adipocyte factor (Fiaf), which inhibits LPL. Since this protein inhibits an inhibitor, increasing Fiaf causes fat cells to release their fatty acids and grow smaller.

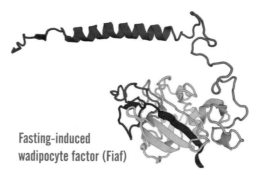

Fasting-induced wadipocyte factor (Fiaf)

Making matters one step more complicated, the microbes in the mouse gut were found to cause less Fiaf production, inhibiting the inhibitor of an inhibitor. Mice with microbes can grow their fat cells better than mice without microbes. The fat cells were not increasing in number; they were just getting fatter. Moreover, when fed a diet high in fat and sugars, colonized mice don't burn fatty acids as well as germ-free mice, due to low Fiaf levels (Bäckhed et al. 2004).

As the previously germ-free mice gained fat, their levels of leptin rose accordingly as expected, since fat cells produce leptin. They also became insulin resistant (Bäckhed et al. 2004). Their livers produced more fat in response to rising glucose and insulin levels. The number of capillaries in their intestines doubled, to the same amount as normal mice have, allowing them to absorb simple sugars at twice their previous rate.

Germ-free mice don't get as fat as normal mice when fed a "Western" diet, which is 41 percent fat, 41 percent sugar and other easily digested carbohydrates (e.g., wheat flour), and 18 percent protein. But when they are given microbes from normal mice, they become obese on this diet (as do normal mice). When both groups of mice ate the same amount of the Western diet, only those with microbes in their gut became obese.

In any symbiosis, it is important for the host to control the hosted organisms. Changes in symbiosis that harm the host have a name: dysbiosis. When we eat foods that change our gut bacteria in nonoptimal ways, the body responds with inflammation, as it would with an infection. The inflammation is at a lower level than with a true infection, but if the bad diet continues, we have chronic low-level inflammation, which causes obesity, but also worsens nonalcoholic fatty liver disease, which is caused by eating sugar. The inflammation then encourages the growth of cancer cells, and we get liver cancer (Bourzac 2014).

Fat does not dissolve in water. In order to digest and absorb fat, our bodies produce a type of detergent that breaks down globules of fat into tiny droplets that can be absorbed. This detergent is called bile acid.

When the diet is low in fiber and high in fat, the Firmicutes outcompete the Bacteroidetes. Among the microbes that bloom are those that digest bile acid into toxic molecules called secondary bile acids. One of these is deoxycholic acid (DCA). There is always a small amount of DCA in the gut, and it seems to have little effect in small doses. But it can damage DNA and cause cancer. It is not only toxic to human cells but also toxic to many of the other microbes in

Deoxycholic acid (DCA)

the gut, the ones that don't produce DCA. This causes a positive feedback loop, where the increased DCA-producing microbes kill off the others and take their place, resulting in more DCA.

The microbes that produce DCA tend to be in the Firmicutes phylum. This explains why obese people (and mice) have more Firmicutes and less Bacteroidetes. But it also partially explains how putting microbes from fat mice into germ-free or normal mice can make them fat. The new microbes are producing toxins that kill off the competition, and they take over, causing inflammation that causes obesity. At the same time, they are better at converting fiber into sugars and fatty acids, so the host gets more calories from a meal.

In one study (Ridaura et al. 2013), researchers worked with human twins where one of the twins was obese and the other was not. They took gut microbes from each of them and transplanted them into germ-free mice. The mice that got microbes from the lean twin produced more short-chain fatty acids (from fermenting fiber). The mice that got microbes from the obese twin had increased metabolism of branched-chain amino acids. Branched-chain amino acid (BCAA) metabolites are more commonly found in obese people than in lean people and are correlated with insulin resistance (Newgard et al. 2009) when combined with a diet high in fat.

When the researchers housed the two mice groups together, all the mice ended up with the lean-type microbes, indicating that on a normal diet they would outcompete the obese-type microbes. But when fed a low-fiber, high-fat diet, the obese-type microbes won out.

In the mice getting microbes from the lean twin, there was extra butyrate and propionate production, due to the fermentation of dietary fiber, and levels of sugars were reduced. These mice had a greater ability to break down fiber.

Just as fiber can protect against the problems caused by fructose, fiber allows us to safely eat more fat, by preventing the fat-eating microbes from outcompeting the fiber-eating ones.

Some 6 percent of the bacteria in the gut are from a single species, *Bacteroides thetaiotaomicron* (BT). BT is one of the bacteria that is good at breaking down fiber into simple sugars and short-chain fatty acids. It needs another microbe, *Methanobrevibacter smithii* (MS), to work most effectively. As its name implies, MS produces methane, the gas that causes embarrassment after eating legumes. This is important, because to produce methane, it consumes carbon dioxide and hydrogen. Since too much hydrogen interferes with BT's ability to ferment fiber, the combination of the two organisms is more effective than either one alone.

When germ-free mice were given just these two organisms, their levels of Fiaf dropped, and they gained fat mass and weight. Given just one organism or the other, they did not gain weight (Bäckhed et al. 2004).

Note that this weight gain was not obesity—the mice simply gained the same amount of fat as normal mice. To become obese, they had to be fed the so-called Western diet, high in sugar, fat, and easily digested starch. The Western diet is characterized by low levels of fiber, so the contribution to obesity from BT and MS would be expected to be small. Indeed, when mice or people become obese, the Firmicutes bacteria grow in number, while the Bacteroidetes bacteria, like BT, diminish.

Microbes in the gut alter the expression of genes in the host. BT causes the human intestine to absorb glucose more effectively. It activates genes to break down fats and transport them (Zocco et al. 2007).

When there is less diversity in the gut microbes, we see an increase in inflammation. In studies of human obesity, low species-richness people had more inflammation (Cotillard et al. 2013). People with fewer species were also fatter and had more insulin resistance and higher cholesterol (Chatelier et al. 2013). It was noted that the association between microbial genes and obesity was greater than the association between human genes and obesity. More of our tendency to be fat is explained by our microbe genes than our human genes. (It was also noted that this study failed to find an increase in Firmicutes or a decrease in Bacteroidetes in obese subjects compared to lean subjects, unlike previous studies.)

The shift away from Bacteroidetes and toward Firmicutes was also found in studies of mice fed a diet high in animal fat. Within the Bacteroidetes phylum, the genus Bacteroides was higher in animal fat diets, and the genus Prevotella was higher in fiber-rich diets (Wu et al. 2011). In a hospital-controlled feeding study, a shift in diet from high fat, low fiber to low fat, high fiber caused a detectable shift from Bacteroides to Prevotella in a mere 24 hours. This was faster than the transit time of the meals through the body—two to four days for the high fiber, and two to seven days for the low fiber (Wu et al. 2011). Nevertheless, the complete shift from Bacteroides-dominated to Prevotella-dominated regimes takes longer than 10 days.

The fiber-rich diet associated with lean body types produces a gut microbiome where Firmicutes is reduced, Bacteroides is reduced, and Prevotella is increased. The obesity-related Western diet creates a gut where Firmicutes and Bacteroides are increased and Prevotella is reduced (Nguyen et al. 2015). Prevotella and Bacteroides are antagonistic—increases in one lead to decreases in the other. Bacteroides is seen predominating in subjects with a diet high in fat and protein. Prevotella is associated with a diet high in plant fiber (Prados 2016).

Looking at the genes of gut microbes, we can see some interesting correlations with aspects of their human hosts. For example, older people have more microbes with genes that digest starch. As humans age, they become less able to digest starch on their own. The starch then reaches the gut, where the microbes break it down (Arumagam et al. 2011). Whether this explains Grandpa's flatulence was not discussed in the paper. More to the point of this book, the host BMI correlates strongly with microbe genes that deal with energy metabolism, supporting the link between obesity and the microbiome.

Diets high in fiber feed the cells of the intestinal lining. The major source of food for these cells is butyrate, the short-chain fatty acid produced when microbes digest fiber. This strengthens the gut barrier against harmful inflammatory molecules in the large intestine.

When fiber is not available, and the diet contains lots of sugars and fat, harmful microbes outcompete the fiber-digesting microbes. These harmful microbes trigger an immune reaction by the host (you). In order to allow immune cells through the gut barrier to fight these microbes, the host makes the gut barrier leaky. Unfortunately, this can allow the inflammatory molecules to get into the bloodstream, and thus the rest of the body. Now we have systemic inflammation—inflammation throughout the system—which leads to obesity, insulin resistance, and type 2 diabetes (Sanz et al. 2015).

The dysbiosis caused by high-sugar and high-fat diets also promotes the growth of bacteria that produce toxic hydrogen sulfide (H_2S), the rotten egg–scented gas. We have enzymes that oxidize this harmful molecule before it can do serious damage to blood and brain cells, but they take time to work. In the gut where the gas is produced, it can damage beneficial microbes and immune cells, as well as the cells of the intestinal lining (enterocytes), and help to cause a leaky gut. Hydrogen sulfide is produced deliberately by some cells in the body because it is a signaling molecule, dilating small capillaries and having some anti-inflammatory effects. But the larger quantities produced during dysbiosis are harmful.

Beneficial bacteria found in the gut of lean test subjects include the following:

- *Bacteroides uniformis* (Sanz et al. 2015)
- *Bacteroides caccae* (Sanz et al. 2015)
- *Bacteroides cellulosilyticus* (Sanz et al. 2015)
- *Bacteroides thetaiotaomicron* (Vrieze et al. 2013)
- *Akkermansia muciniphila* (Dao et al. 2014)
- *Faecalibacterium prausnitzii* (Vrieze et al. 2013)
- *Bifidobacterium bifidum* (Sanz et al. 2015)
- *Alistipes finegoldii* (Sanz et al. 2015)
- *Lactobacillus casei* (Sanz et al. 2015)
- *Prevotella copri* (Prados 2016; Ley 2016; Kovatcheva-Datchary et al. 2015)
- *Methanobrevibacter smithii* (Goodrich et al. 2014)
- *Christensenella minuta* (Goodrich et al. 2014)

Harmful bacteria associated with obesity are:

- *Staphylococcus aureus* (Sanz et al. 2015)
- *Clostridium leptum* (Sanz et al. 2015)
- *Eubacterium hallii* (Sanz et al. 2015)
- *Escherichia coli* (Sanz et al. 2015)

The idea of curing microbial diseases in the large intestine by transplanting bacteria from healthy individuals was first tried in 1958 (Eiseman et al. 1958). A disease caused by antibiotic use, pseudomembranous colitis, was a bacterial infection of the large intestine. It was cured by fecal enemas, which restored a healthy ecology to the patient's gut.

Recently, with the new understanding of how the gut microbiome affects obesity and diabetes, studies have been done on transplanting fecal matter from healthy lean individuals into the gut of obese patients (Vrieze et al. 2012). One study found improved insulin resistance in the test subjects after the procedure. The success of studies like this have led to proposals to select a variety of pure microbe cultures to encapsulate for oral administration, both to reduce the "ick" factor and to eliminate the risks involved in administering natural "cultures" that might contain unknown harmful species. However, as the ecology of a healthy gut is quite complex, and some microbes are difficult or impossible to grow as pure cultures, such artificial colonies may take some time to develop.

Human genetics can affect which bacteria end up colonizing the human gut (Goodrich et al. 2014). In a study of twins in the UK, comparing identical twins (monozygotic, or MZ) to fraternal twins (dizygotic, or DZ), researchers were able to tease apart the contributions of genetics from those of a shared environment and a shared mother, from whom gut bacteria are transferred at birth.

The microbiomes of the MZ subjects were more similar than those of the DZ subjects. The differences in Bacteroidetes were mostly due to environmental factors. This correlates with observations that Bacteroidetes responds well to diet interventions. The differences in *Ruminococcaceae* (Firmicutes) and *Lachnospiraceae* had the strongest genetic component of heritability. Unsurprisingly, the types that were most affected by genetics proved to be the most stable over time—the environment can change, but genetics typically do not.

The family most affected by genetics was *Christensenellaceae* (a Firmicutes). Forty percent of the difference in this family is due to human genetics. This family turned out to be a hub of a group of microbe types that occur together, and includes *Methanobacteriaceae* (an *Archaea*) and *Dehalobacteriaceae* (a Firmicutes). When this group of microbes is present, there are fewer *Bacteroidaceae* and *Bifidobacteriaceae*. This is important because the *Christensenellaceae* group is associated with low BMI (Goodrich et al. 2014). Other groups of microbes associated with low BMI are *Oscillospira* and *Methanobrevibacter smithii*. Interestingly, *Christensenellaceae* is enriched in omnivorous mammals but not in herbivorous or carnivorous ones.

The association of *Christensenellaceae* with *Methanobrevibacter* was important for the association with low BMI. One without the other had less effect than the two together (Goodrich et al. 2014). The methanogens produced higher levels of the fatty acids propionate and butyrate than microbiomes lacking methanogens. On the other hand, methanogens were not required to reduce obesity—the *Christensenellaceae* alone was sufficient to reduce body fat. Merely adding cultures of *Christensenella minuta* to obese mice caused them to lose weight.

Adding the *Christensenellaceae* remodeled the diversity of the mouse gut microbiome, causing rises in *Oscillospira*. Because *Christensenellaceae* was still found only in low levels, its effects on the rest of the gut microbiome may indicate that it is a key species, able to have effects beyond its own contributions (Goodrich et al. 2014).

The diversity of the gut microbiome is important. It isn't just a few good or bad microbe types. They all work together as an ecology. When we create obese mice by feeding them a high-fat, high-sugar, low-fiber diet, the 50 percent reduction in Bacteroidetes and compensating increase in Firmicutes happens mostly through reductions in certain species, leaving the overall diversity much lower (Ley et al. 2005). A diet of exclusively McDonald's fast food can cause a large reduction in diversity, this time by raising Bacteroidetes and lowering Firmicutes, due to the high protein and fat in that diet (Spector 2015). Despite having the opposite ratios of the two classes of microbes, this diet also led to dysbiosis, obesity, and inflammation.

Antibiotics reduce gut microbe diversity, as they kill some microbes more easily than others. In the absence of antibiotic use, our gut diversity generally grows as we age (Z. Xu and Knight 2015). The more different types of plants we eat, the more our gut diversity grows. Some microbes need food that they can get only when we eat their favorite plants. Eating a wide variety of plants means that all the microbes get to eat their preferred foods. Modest alcohol use, at the level of one drink per week, also raises gut microbe diversity.

As we saw with *Christensenellaceae*, some species can change the ecology for the better, while remaining in low percentages themselves. Besides

Christensenellaceae, the yeast *Saccharomyces boulardii* also improves microbial diversity in the gut, by providing food for beneficial bacterial species (Everard et al. 2014). There are likely to be many more of these keystone species found as we continue to learn more about our symbiotic gut microbes.

PROBIOTICS

The human gut has evolved to enable us to be omnivores. The upper gut is almost sterile, like that of a carnivore, adapted to eating fat and meat. The lower gut (the large intestine) teams with bacteria that help us digest fiber and vegetables, similar to the gut of horses and cows. As a species, we are able to eat a diet like that of Yupiks and Inuit in the far north, who live on fat and protein, or a diet like that of a pure vegan, eating no animal products at all.

However, when we learned to farm and to cook food, our diet changed. New foods that require cooking, such as grains, legumes, and root vegetables, came into our diet. These foods need to be cooked in order for us to digest them. Try eating a dried bean or a raw potato or a mouthful of wheat kernels. Cooking and farming were great inventions that allowed humans to access much more food and to expand to cover the planet.

Unfortunately, these new foods are full of starches that can be fermented in the normally sterile upper gut. The stomach, which is acidic to digest protein, usually kills the bacteria that are sensitive to acid. The next organ in the digestive process, the duodenum, is alkaline, and kills bacteria that are sensitive to alkali. Bile salts are toxic to most microbes, and they are added in this phase of digestion. Adding a lot of cooked starches and sugars allows bacteria to colonize farther down in the small intestine, where they can cause inflammation and allergic reactions.

The large intestine is where anaerobic bacteria—bacteria that die in the presence of oxygen—such as Bacteroides can thrive. These bacteria ferment the fiber and resistant starch that the human digestive system can't handle. They produce the short-chain fatty acids that are needed by the cells in the intestinal wall, and they contribute as much as 500 extra daily calories to the diet. The fermentation also produces heat to warm our bodies.

Probiotics are cultures of bacteria known to be beneficial to humans. They do not necessarily have weight loss or weight maintenance as their primary function, but they aid in gut microbiome health.

Many fermented foods are excellent probiotics. Yogurt, kefir, sauerkraut, pickles, natto, kombucha, kimchi, tempeh, and lassi are some examples. Yogurt that contains sugar or fruit is not effective, as the sugar inhibits the growth of the bacteria.

While natural fermented foods are great, there are drawbacks. Many of the bacteria cannot survive the trip through the digestive system. The acidic stomach and the alkaline duodenum are both detrimental to most bacteria (this is generally a good thing, as many bacteria are harmful). Also, the fermented foods are limited in how many bacteria are available. Concentrated dried cultures of probiotic bacteria in capsule form overcome these limitations—they can contain 50 or more times as many bacteria, and they come encapsulated in protective coatings that can survive until they reach the large intestine. The drawback is that they generally contain only a few of the beneficial species we want to cultivate.

Some typical species found in probiotic formulas:

- *Bifidobacterium bifidum*
- *Bifidobacterium lactis*
- *Bifidobacterium longum*
- *Bifidobacterium breve* (Coakley et al. 2002)
- *Lactobacillus plantarum* (Vrieze et al. 2013)
- *Lactobacillus casei* (Vrieze et al. 2013)
- *Lactobacillus acidophilus* (Vrieze et al. 2013)
- *Lactobacillus gasseri*
- *Lactobacillus salivarius*
- *Lactobacillus rhamnosus*
- *Lactobacillus bulgaricus*
- *Lactobacillus paracasei*
- *Saccharomyces boulardii Biocodex*

Bifidobacterium bifidum produces acetic acid, which, as we have seen, is a fatty acid with important effects in metabolism and satiety. It also converts linoleic acid (LA) into conjugated linoleic acid (CLA) (Coakley et al. 2002), which has many health benefits including weight loss (see the section on dietary fat). *Bifidobacterium breve* and *Bifidobacterium dentium* have been found to be the best at this conversion. *Bifidobacterium pseudocatenulatum* is also a good producer of CLA (Raimondi et al. 2016).

Lactobacillus plantarum and *Lactobacillus curvatus* have been shown to reduce weight gain in mice fed a high-sugar, high-fat diet (Yoo et al. 2013). Fat accumulation both in fat tissue and in the liver were significantly reduced, especially when both species were present. The study points out that multistrain probiotics are more beneficial than single-strain probiotics.

Lactobacillus gasseri reduced abdominal fat and subcutaneous fat by almost 5 percent in adult obese human subjects in a randomized control trial (Kadooka et al. 2010). Taken in the form of milk fermented with the bacterium (yogurt), it also reduced cholesterol levels.

Another plant-derived lactic acid bacterium, *Pediococcus pentosaceus*, was also found to be effective in preventing obesity in mice fed a high-fat diet (X. Zhao et al. 2012). Fat deposits in the liver were also reduced. Similar results were found with the bacterium *Bacteroides uniformis* (Cano et al. 2012).

Saccharomyces boulardii is one of the few yeasts known to be a probiotic associated with reduced body fat, liver fat, and inflammation (Everard et al. 2014). It may work more through changing the gut ecology and promoting the other antiobesity microbes than through specific actions on its own. Unlike other yeasts, it prefers body temperature, can withstand the acidic environment of the stomach, and is resistant to bile acids. These features allow oral doses to survive until they can colonize the large intestine. Oral dosages of *S. boulardii* raise Bacteroidetes by 37 percent, and decreased Firmicutes by 30 percent compared to controls. No changes in food intake were seen, despite a drop in weight and total fat mass of 10 percent. Markers of inflammation dropped by 40 to 50 percent.

You may have noticed that the common strains found in probiotic capsules are limited to only *Lactobacillus* and *Bifidobacterium*. These are easy-to-culture, gram-positive, anaerobic bacteria. Finding a probiotic that has *Bacteroides* or *Pediococcus* will probably be difficult. To increase these strains in your gut, you will want to turn to prebiotics to enhance the numbers that already exist there.

PREBIOTICS

A prebiotic is a fiber that feeds gut microbes. Certain types of fiber are known to promote the growth of the kind of gut microbes that are beneficial for weight control, insulin sensitivity, and reducing metabolic syndrome. They are better at preventing obesity than at curing it, but combined with successful weight-loss efforts, they will help prevent regaining the weight. Because they are fiber, they are filling and kill appetite, making dieting easier.

High-sugar, high-fat diets change the gut microbiome, specifically by reducing *Bifidobacteria* (Vrieze et al. 2013). Supplementing the diet with the fiber oligofructose was found to completely restore levels of *Bifidobacteria*, leading to improved glucose tolerance, increased satiety, and weight loss. The poor diet had increased the levels of bacteria that contain lipopolysaccharides, which cause inflammation. (The name of the molecule tells us that it is a mixture of fatty acids and chains of sugars.) Lipopolysaccharides form part of the cell wall of certain bacteria and are called endotoxins, toxins that are released only when the bacterial wall is broken. Endotoxins provoke a strong response from the human immune system, causing inflammation, fever, and diarrhea, and sometimes potentially fatal septic shock.

Eating a low-carbohydrate, high-protein diet, often recommended for weight loss, has harmful effects on gut bacteria (Duncan et al. 2006), due to the lack of food for the microbes. This reduces the proportion of important *Bacteroides* species and *Roseburia*, both of which are important for weight management. A diet that includes fiber and resistant starch can keep these beneficial strains in good health while still reducing the levels of digestible carbohydrates.

Oligofructose is a molecule made up of 10 or fewer fructose molecules joined together. It is not digested by humans, so the fructose does no harm. Nevertheless, it tastes somewhat sweet—a third to half as sweet as sugar—and is used in some syrups as a noncaloric sweetener. Commercially processed oligofructose powder normally contains 6 to 10 percent sugars: glucose, fructose, and sucrose. Specially purified versions with longer chains of 25 to 60 fructose molecules do not taste sweet, have no sugars, and are used as fat-mimicking fiber additives to some processed food.

Natural sources of inulin, the longer chain version of oligofructose, are jicama, Jerusalem artichokes (also known as sunchokes), chicory root, burdock root, dandelion root, salsify root, and yacon root.

Less concentrated sources of inulin are asparagus, leeks, onions, garlic, plantain, and bananas.

If your body has not been getting lots of fiber, and you have the microbiome of an obese person, it might take a little while to grow the lean-type microbes in your gut. Taking in too much inulin or oligofructose before your gut microbiome has been conditioned can cause uncomfortable gas and bloating. Slowly increase the amounts of inulin in your diet and back off a little if you feel too gassy. Over time you will grow a new microbial garden that handles the inulin perfectly.

ANTIBIOTICS

Antibiotics are generally not very specific in which microbes they kill. This means that in the human gut, they kill the beneficial microbes as well as any harmful ones. As a consequence, they can alter the state of the microbial ecology in the gut in ways that affect metabolism and obesity.

If you have obese-type microbes, you might see improved glucose tolerance and insulin sensitivity after a dose of antibiotics. The effect will be short lived if your diet does not change to favor lean-type microbes.

Antibiotics may also help obese patients by reducing low-grade inflammation. Again, since this is caused by the microbes that thrive in a high-sugar, high-fat, low-fiber diet, the effect will not be sustained unless the diet changes to a higher-fiber, lower-fat, low-sugar diet. Antibiotics reduce lipopolysaccharides, since those are produced by harmful bacteria. This reduces inflammation and reduces the harmful metabolic effects of lipopolysaccharides.

Antibiotics also reduce short-chain fatty acids in the large intestine, since these are produced by beneficial microbes. This can result in fewer calories being absorbed in that organ.

Various microbes have different tolerances for different antibiotics. This is almost guaranteed to cause problems in the gut by upsetting the ecology of the gut and reducing the diversity of the microbiome. As we have seen, a less diverse microbiome leads to obesity, and many microbes require the assistance of other species of microbe in order to thrive.

The rapid rise of the obesity epidemic since the 1970s may be caused in part by increased use of antibiotics. The reduced diversity of the gut microbiome leads to higher inflammation, insulin resistance, and higher blood lipids and cholesterol.

If you must take antibiotics for some condition, take them along with probiotics and prebiotics that help the beneficial microbes withstand the antibiotic and recover faster once the antibiotic is discontinued. This will help to create the proper gut ecology, one that does not promote obesity.

GENETICS

As we have seen in the section on hormones, there are lots of "moving parts" to the obesity problem. All those hormones and neurotransmitters, and their receptors, are produced according to genetic instructions in your DNA. We know that genes are inherently variable—without variation, there can be no evolution. Some variants of genes work better than others. If you think about what a random mutation is, you can see that it is unlikely to improve on something that has had millions of years of selection pressure to create. It is like throwing a rock at a running car engine and hoping for an improvement. Most likely, the engine won't run quite as well afterward.

We know we can't blame genetics for the obesity crisis, because our genes did not change all of a sudden in the late 1970s, when we all started getting fat. On the other hand, we all share the same food-rich environment, and yet some of us are fat, while others are less so, or lean. That variation may well be due to genetic effects, since we know that obesity is heritable—fat runs in families. Somewhere between 40 and 70 percent of people's BMI can be blamed on their parents' genes, although some papers claim 50 to 90 percent (Barsh, Farooqi, and O'Rahilly 2000). Adopted children have BMIs that correlate more with their birth parents than with their adoptive parents. When sets of twins are overfed, the weight gain is similar in the identical twins but differs in fraternal pairs (Ramachandrappa and Farooqi 2011).

If something goes wrong genetically with the complex systems that maintain your body fat balance, we can sometimes see dramatic effects. Scientists study morbidly obese people—100 pounds above ideal weight, BMI over 40, significant medical problems due to obesity—to see if there is a genetic cause for these outliers. This is how we learn a lot about the genetics of obesity—by studying the extremes.

But there are far more examples of subtle variations. These don't automatically make a person destined to be 600 pounds. They just make a person more likely to gain weight than someone without that variation. There are many hormones and enzymes and receptors involved in energy balance that, combined with natural variation in all genes, allow for thousands of tiny effects. It is these effects as a whole that combine to give a probability of becoming overweight or obese given a food-rich environment and a food-obsessed social setting.

HEDONIC EATING

An example is what we call hedonic eating. Eating for pleasure. We all do it. But some people with particular variations in the genes for the brain messenger molecules dopamine and opioids get more pleasure from food than other people do. In studies of binge-eating disorders, it was found that two genes in particular (called DRD2 and OPRM1) had what is called gain-of-function variations in some people, and loss-of-function variations in others. The people who had gain-of-function in both genes were prone to binge eating. The people with loss-of-function in both genes were less likely to be in the binge-eating group (Davis et al. 2009).

This does not mean that these people are destined to be obese. It just means that they are more likely to overindulge when foods rich in sugar, salt, and fat are around. They can avoid situations like that, or fight the temptations, once they know what their triggers are.

The OPRM1 gene codes for a receptor for opioids. Gain-of-function variations in that gene have been tied to the reward baby monkeys get from attachment to their mothers, to increased alcohol stimulation in humans, and to a greater tendency toward drug use and abuse.

On the other end of the spectrum, the dopamine receptor gene DRD2 leads to obesity by a different mechanism. Obese people who are *not* binge eaters tend to have a loss-of-function variation in this gene. They are less able to experience natural reward. This is linked to alcoholism and gambling, where those with this variation seek more stimulation to make up for the lack of reward. Eating lots of highly palatable foods full of sugar and fat is another way to self-medicate to get more reward.

INFLAMMATION

Your genetic makeup can make you more prone to inflammation, which can lead to obesity. Mutations in genes relating to TNF-alpha and IL-6 (G-308A and C-124G) in obese patients with impaired glucose tolerance carry twice the risk of type 2 diabetes (Dandona, Aljada, and Bandyopadhyay 2004).

CARBOHYDRATES

The DioGenes large-scale randomized controlled test (Astrup, Raben, and Geiker 2015) studied high-protein, low glycemic load diets and reported that a genetic variant in the gene rs987237 was strongly associated with how well the diet helped. Sixty-seven percent of the subjects had the AA variant and did well on the high-protein, low glycemic load diet. The other third of the subjects, with either the AG or the GG variants, saw no differences in weight whether they were on the high-protein, low glycemic load or the normal-protein, high glycemic load diets. The G variant is associated with obesity, and having it on one half or both halves of the DNA strand was enough to lose the benefit of the high-protein diet. If you are in the unlucky one-third, the Atkins and other low-carbohydrate diets will have less effect on you.

THE OB GENE

Defects in the famous ob gene, so named because it causes obesity in mice, are rare in mice and also quite rare in humans (Carlsson et al. 1997). But if both copies of the gene are damaged, uncontrolled feeding occurs, since the leptin receptor no longer works and the unfortunate animal (or person) is always hungry. Up to 3 percent of patients with severe obesity have loss-of-function defects in the LEP gene, which codes for leptin (Ramachandrappa and Farooqi 2011). These leptin-deficient individuals also have impaired fat burning, further exacerbating their condition. Worse yet, leptin is needed for proper thyroid function, and thyroid hormone controls metabolic rate.

THE FTO GENE

In a wonderful study of the genetics of obesity (Claussnitzer et al. 2015), researchers looked into the area of the genome that harbors the strongest association with obesity, an area known as the FTO region. FTO stands for "FaT mass and Obesity-associated," and the region resides on chromosome 16.

It turns out there are several variations in the FTO area that have an effect on obesity and energy balance. Some affect the hypothalamus and influence snacking. But the region in this particular study, called rs1421085, determines whether or not an adipocyte progenitor cell turns into beige fat, which burns energy to create body heat, or white fat, which stores triglycerides. The study looked at what is called a single nucleotide polymorphism (SNP, pronounced *snip*), where one of the DNA bases is changed from a thymine (T) into a cytosine (C).

This region is a regulatory gene (one that controls other genes, rather than coding for a protein). Moreover, the gene it regulates is yet another regulatory gene, ARID5B, which in turn regulates two genes, IRX3 and IRX5, that finally do code for proteins.

IRX3 and IRX5 are genes that make the adipocyte progenitor cell turn into a white fat cell. ARID5B turns those genes on ("upregulates" them).

The rs1421085 SNP downregulates this enhancer. When the T is changed to a C, this inhibiting function is lost, and the ARID5B gene tells the progenitor cells to all turn into white fat cells instead of heat-producing beige fat cells.

Without enough beige fat, less energy goes into generating heat, and there are more white fat cells to take up that unburned fuel and store it. People with the working form of the gene are protected from much diet-induced obesity.

This SNP accounts for 1 percent of the variance in BMI in the population, and 22 percent of the population-attributable risk of obesity. This means that there are plenty of people with the mutation who are not obese or overweight, but among people who are obese, this SNP accounts for over a fifth of the likelihood.

The rs1421085 SNP is both widespread and has strong effects on those carrying the variation. In Europeans, 44 percent of the population carry the variant. In African populations, only 5 percent carry it. This suggests that either it was selected for (has some survival advantage) in European populations, or that there was a genetic bottleneck, where by chance only those with this variant survived. The T variant—the version that works—is strongly conserved, meaning that it occurs in animals ranging from mice and rats to horses and cattle, indicating that it is very important for survival. In humans, it may not have been as important once we invented clothing and discovered fire, since its effect is to keep us warm. The shift from energy-dissipating tissue (beige fat) to energy-storing tissue may thus reflect that energy storage was more important for European populations in the distant past than thermogenesis.

The SNPs in the FTO region are usually coinherited, so any one person might have several obesity-related SNPs that combine to make it more likely that they will become obese in our new environment of highly available obesogenic foods.

We saw in the section on hormones that many different systems have effects on obesity. Cortisol, the stress hormone, and CRH (corticotropin-releasing hormone) link stress to obesity. MCH (melanin-concentrating hormone) and alpha-MSH (alpha-melanocyte-stimulating hormone) link sleep and pigmentation to obesity. Endogenous opioids (endorphins and enkephalins) link obesity to the reward systems in the brain.

It turns out that all these hormones are produced by splitting up a single protein produced primarily by the pituitary gland. This protein is called pre-pro-opiomelanocortin (pre-POMC). As the name implies, this protein controls the opioid, melanin, and corticotropin hormones, as well as the lipotropins, hormones that promote the release of fat from the liver into the blood. The protein is also produced in the brain's hypothalamus, in the immune system, and in the skin (remember it is the source of hormones that control pigmentation) (Pritchard, Turnbull, and White 2002).

It is no surprise then that defects in the gene that codes for the POMC protein can result in obesity. In the general case of POMC deficiency, where the whole protein is missing or very short, not enough adrenal hormones are produced, the appetite-depressing hormone alpha-MSH is not produced, and the melanin hormones are not produced. Babies with this defect have light skin and red hair, are ravenously hungry, and experience frequent periods of low blood sugar (sometimes leading to seizures), elevated levels of toxic bilirubin in the blood, and problems producing the bile acids that allow fats to be digested. The good news is that only about 50 cases have ever been described. This is not surprising, as this protein is vital to survival, and evolution would quickly weed out carriers of the defect if the defect occurs in both copies of the gene, one from the mother and one from the father.

People who have one working copy of the gene produce less of the POMC protein and have an increased risk of obesity as well as other problems. About 2 percent of all severe early-onset obesity is caused by having only one working POMC gene.

The POMC gene is also what is known as a quantitative trait locus (QTL) for obesity. This means that slight changes in it affect many other systems that slightly increase the odds of being obese. Small changes in the POMC region are linked to low leptin levels in the blood, thus decreasing the appetite suppression that leptin provides.

Other SNPs in the FTO gene, which some people call the fat gene or the fatso gene, are rs9939609 (linked to type 2 diabetes), rs7193144, rs9940128, and rs1121980 (early-onset obesity), rs9939973 (obesity), rs8050136, rs10938397, and rs2815752 (diabetes and obesity), rs7498665, rs4752856, rs17782313 (high BMI, more snacking). Note that some of these variations affect beige fat and fat burning, without appetite changes, while others (like rs17782313) affect appetite and feeding frequency (snacking).

THE TUB GENE

Another gene with a cutesy name is TUB (or "tubby"). There are three SNPs in that gene that are of interest with respect to obesity.

The SNP rs1528133 is associated with diabetes and high BMI and appears to affect carbohydrate metabolism. People with the AA type are considered "normal." People with the AC or CC variants have higher BMI numbers than the AA people do.

The SNP rs2272382 affects fat metabolism. People with the AA or AG variant get more energy from saturated fat and monounsaturated fat than those with the GG variant. The AA and AG types also tend to eat more sugar (Vliet-Ostaptchouk et al. 2008). The SNP rs2272383 was similar to rs2272382, but with a weaker effect.

If you have any of these risk types, you are prone to eating more sugar. Knowing that this is a risk factor for you might help you to cut down on or eliminate sugars from your diet. Of course, as we have seen, eliminating sugar is beneficial to all genotypes, not just those genetically predisposed to overdoing it. When diets turned away from fat and toward more carbohydrate (especially sugars) in the late 1970s, people with nonfunctional TUB genes were likely to be affected more than others.

THE MC4R GENE

The alpha-melanocyte-stimulating hormone (alpha-MSH) produced from POMC acts on one of the five known melanocortin receptors in the brain, called MC4R (melanocortin-4 receptor). This receptor is specific for appetite and feeding behavior. Mice with the MC4R receptor gene knocked out are constantly eating and become quite obese. One form of morbid human obesity has been shown to be caused by a mutation in this gene (Vaisse et al. 1998). Mutations in this one gene are the most common obesity mutations known so far. Even people with one working copy (heterozygous) of the gene have increased risk of obesity. There are almost 200 variations of this gene in different populations—Asian, North American, European—and it accounts for 2 to 3 percent of all genetic obesity.

Mutations in the MC4R gene cause hyperphagia (eating a lot) in children, and this decreases with age. The metabolic rate is unaffected. The receptor is also expressed in bone cells, and the obesity-related mutations cause bone to accumulate more than normal, with less bone being mined for minerals like calcium. Thus, these mutations make the bones heavier. While the mutations increase appetite and obesity, they do not predispose carriers to diabetes like other forms of obesity, and they lower blood pressure.

Depending on which MC4R mutation a person has, the effect can range from a slight reduction in receptor activity to a complete lack of activity. It is possible to have both copies of the gene nonfunctional, which always leads to early-onset obesity, but this is rare. Also rare, but with very high risk of obesity, is to have one reduced-function mutation from one parent and another from the other. With one functional copy and one reduced-function copy, the risk of obesity can range from slight to quite high.

People with reduced MC4R activity benefit more from exercise than people who are obese for other reasons, possibly due to the exercise-produced hormone irisin and its effects on tipping progenitor cells into becoming bone instead of fat, and its effects on raising thermogenesis (production of body heat). People with these mutations in both copies of the gene respond less well to bariatric surgery than those with other forms of obesity or with only one damaged copy.

A study where normal subjects without MC4R mutations were given twice-daily nasal spray containing the alpha-MSH hormone for six weeks showed an average loss of body fat of 3.7 pounds, or a little over half a pound per week (Fehm et al. 2001). Insulin levels dropped by 20 percent and leptin dropped by 24 percent, without increasing appetite.

Another melanocortin receptor, MC3R, is also associated with obesity, but without increasing appetite. The effects of defects in this gene on obesity are small, however, and only add a little extra risk of becoming obese. Tiny effects like this, spread throughout the genome, are what make genetic obesity so difficult to piece out. All of us have hundreds of small differences in key energy-balance genes, and each one contributes just a little bit of risk.

Mice with nonfunctional MC3R receptors have increased fat mass and decreased lean mass and get more calories from their food than their normal littermates. They eat less and have normal metabolic rates, but they gain fat nonetheless.

The hormone adiponectin is another obesity-related molecule. Adiponectin levels are controlled by genes that have variants that affect obesity. Low adiponectin levels are associated with metabolic syndrome: obesity, diabetes, etc. Loss-of-function variants in the genes that code for the hormone or its receptors increase the risk of high BMI and problems in glucose metabolism.

Another molecule, apolipoproteins, are the proteins that bind to fat and cholesterol so they can be soluble in water and carried through the bloodstream. They act like detergents, with one side that is attracted to water and another side that is attracted to fats. Variations in the genes that code for them affect heart disease risk and whether saturated fat leads to obesity. Two genes, APOA2 and APOA5, have variants that are simultaneously linked to obesity and to lower risk of cardiovascular disease.

One important thing to note about genetic causes of obesity: they are not a guarantee that you will be obese. There are many carriers of each of these genes that are not overweight or obese. And no single gene is found in more than 3 to 5 percent of all people with BMI over 40. Before 1980, obesity was an unusual sight. We have changed our environment in ways that combine with preexisting genetic variants to cause us to gain weight. While we cannot yet do much about our genetics, we have a lot of control over our environment.

SUMMARY

The following table shows a list of single nucleotide polymorphisms (SNPs) that have links to obesity, BMI, or diabetes. If you have the data from a genetics screen, such as from 23andMe, you can look up the SNP name in your data to see if you have the normal allele or the risk allele. If you wish to know more about any particular SNP, you can search for the SNP name (e.g., rs17782313) on the Internet and generally get quite a lot of information, including scientific papers that studied that SNP or gene.

Gene	SNP	Allele	Effect
FTO	rs9939609	TT	Lower risk of obesity and diabetes
		AA or AT	Higher risk of obesity and diabetes
	rs1121980	CC	Normal
		CT or TT	High and higher (respectively) risk of obesity
	rs1558902	AA or AT	Higher BMI
		TT	Normal
APOA2	rs5082	CT or TT	Normal
		CC	Saturated fat leads to obesity, but lower heart risk
APOA5	rs662799	AA	Normal
		AG or GG	Less weight gain on high-fat diet, but heart risk
MC4R	rs2229616	GG	Normal
		AA or AG	Lower risk of metabolic syndrome
	rs17782313	TT	Normal
		CT or CC	Higher risk of obesity
	rs571312	AA or AC	Higher risk of obesity
		CC	Normal

Gene	SNP	Allele	Effect
TUB	rs1528133	AA or TT	Normal
		AC or CC	Higher BMI
	rs2272382	AA	Normal
		AG or GG	Higher BMI
	rs2272383	AA	Normal
		AG or GG	Higher BMI
Adiponec-tin levels	rs1851665	GG	Higher adiponectin levels
		AG	Average adiponectin levels
		AA	Lower adiponectin levels (obesity)
	rs7193788	AG	Average adiponectin levels
		AA	Higher adiponectin levels
		GG	Lower adiponectin levels (obesity)
	rs6444175	GG	Average adiponectin levels
		AG	Slightly lower adiponectin levels
		AA	Lower adiponectin levels (obesity)
ADIPOQ	rs864265	GG	Normal
		GT or TT	Higher BMI
	rs266717	TT	Normal
		TC or CC	Higher adiponectin levels
ARL15	rs4311394	AA	Normal
		AG or GG	Lower adiponectin levels (obesity)
MSRA	rs545854	GC or GG	Central adiposity (visceral fat)
		CC	Normal
	rs7826222	CC	Normal
		GG or GC	Higher risk of obesity
LYPLAL1	rs2605100	GA	Central adiposity (visceral fat)
		AA	Normal

Gene	SNP	Allele	Effect
TFAP2B	rs987237	AG or GG	Central adiposity (visceral fat)
		AA	Normal
MTCH2	rs10838738	GA or GG	Higher risk of obesity
		AA	Normal
	rs3817334	TT or TC	Higher risk of obesity
		CC	Normal
NEGR1	rs2815752	AA or AG	Higher risk of obesity
		GG	Normal
TMEM18	rs6548238	CC or CT	Higher risk of obesity
		TT	Normal
	rs2867125	CC or CT	Higher risk of obesity
		TT	Normal
SH2B1	rs7498665	TT or TC	Higher risk of obesity
		CC	Normal
	rs7359397	TT or TC	Higher risk of obesity
		CC	CC
SEC16B	rs10913469	CC or CT	Higher risk of obesity
		TT	Normal
	rs543874	GG or GA	Higher risk of obesity
		AA	Normal
BDNF	rs6265	CC or CT	Higher risk of obesity
		TT	Normal
	rs10767664	AA or AT	Higher risk of obesity
		TT	Normal
LRRN6C	rs10968576	GG or GA	Higher risk of obesity
		AA	Normal
FAIM2	rs7138803	AA or AG	Higher risk of obesity
		GG	Normal
KTCD15	rs29941	GG or GA	Higher risk of obesity
		AA	Normal

Gene	SNP	Allele	Effect
ETV5	rs7647305	CC or CT	Higher risk of obesity
		TT	Normal
	rs9816226	TT or TA	Higher risk of obesity
		AA	Normal
GNPDA2	rs10938397	GG or GA	Higher risk of obesity
		AA	Normal
SEC16B	rs543874	GG or GA	Higher risk of obesity
		AA	Normal
TNNI3K	rs1514175	AA or AG	Higher BMI
		GG	Normal
NUDT3	rs206936	GG or GA	Higher BMI
		AA	Normal
MTIF3	rs4771122	GG or GA	Higher BMI
		AA	Normal
MAP2K5	rs2241423	GG or GA	Higher risk of obesity
		AA	Normal
NRXN3	rs10150332	CC or CT	Higher risk of obesity
		TT	Normal
RBJ	rs713586	CC or CT	Higher risk of obesity
		TT	Normal
GPRC5B	rs12444979	CC or CT	Higher risk of obesity
		TT	Normal
QPCTL	rs2287019	CC or CT	Higher BMI
		TT	Normal
SLC39A8	rs13107325	TT or TC	Higher BMI
		CC	Normal
FLJ35779	rs2112347	TT or TG	Higher BMI
		GG	Normal
TMEM160	rs3810291	AA or AG	Higher BMI
		GG	Normal

Gene	SNP	Allele	Effect
FANCL	rs887912	TT or TC	Higher BMI
		CC	Normal
CADM2	rs13078807	GG or GA	Higher BMI
		AA	Normal
PRKD1	rs11847697	TT or TC	Higher BMI
		CC	Normal
LRP1B	rs2890652	CC or CT	Higher BMI
		TT	Normal
PTBP2	rs1555543	CC or CA	Higher BMI
		AA	Normal
ZNF608	rs4836133	AA or AC	Higher BMI
		CC	Normal
RPL27A	rs4929949	CC or CT	Higher BMI
		TT	Normal

ENVIRONMENT

Although genetics plays a role in obesity, it acts more like a multiplier of environmental effects than as the main cause. In an environment that is full of sugary, fatty, salty foods, everyone gains weight. But those with genetic tendencies toward imbalance get fatter sooner. If both types of people are fed a diet that is more natural, without sugar, wheat flour, and processed foods, even those with obese-type genes won't have problems.

In a large study of 20,000 Scottish families, genetic factors (identified using both gene sequencing and pedigree) accounted for only half of traits relating to obesity and heart health (Xia et al. 2016). What the study called "recently-shared environment of couples" accounted for approximately 11 percent of the differences. It seems that whom you marry accounts for a significant proportion of your problem, or lack of one. It also means that if your spouse has a weight problem, *you* might be it.

Environment is not just the foods you bring home or the people you live with. Dieting at a party is difficult for several reasons. First, the host wants to please everyone, so the foods are there not to fend off hunger but to stimulate eating for pleasure and reward. They are sweet, rich, and often salty. They trigger the opiate receptors in our brains, and the sight of them stimulates the dopamine that makes us want the food.

Second, there is usually a large variety of foods. We have evolved to desire variety, since it helps to get all the different types of nutrients we need. Given a single type of food, we will eat it for a while and then tire of it. Given a variety, we will keep eating, trying one food after another. If the foods are all mixtures of cheap sugar, salt, and fat, we won't get the nutrients, but our brain doesn't know that.

Third, there are lots of other people eating. Seeing others eat makes us want to eat. It doesn't help that there are a bunch of people at the party telling us we absolutely *must* try the cheesecake, or that they brought their favorite new dessert and you will just die if you don't get a slice.

Fourth, when we are in a place where there are lots of people and lots of food, we will mindlessly keep eating, not paying any attention to whether we are hungry or stuffed to the gills. We talk, and we reach for a snack in between breaths.

In a restaurant, dieting is difficult for many of those reasons, but also for a few more. Restaurants know that people associate large portions with getting their money's worth. They serve a whole day's worth of calories on one plate. As long as the food is there in front of us, and we are enjoying the company and the conversation, we will mindlessly clean our plates and complain afterward about the bloated feeling we have in our stomachs. Instead of eating less, we buy drugs that reduce stomach acid, thinking that heartburn is just something we are prone to getting when we eat, not a signal that we have overindulged.

At work, there may be pastries or doughnuts at meetings, candy at the reception desk, or a vending machine in the hallway. Lunches may be in the company cafeteria or at a restaurant, where variety and processed dessert foods conspire to defeat our willpower.

At the supermarket, the processed foods take up nearly the whole store. The produce, meat, and dairy sections are all on the periphery. Inside is row after row of foods made of wheat starch, sugar, cheap processed vegetable oils, and salt.

On the drive home are many fast food restaurants and ministores beckoning us in. Just the sight of them stimulates our cravings and raises insulin levels. What sugar there was in our blood is diverted to fat. The low blood sugar that results makes us hungry. And it is so easy to just stop in and get a milkshake or a box of cookies and a Big Gulp.

At home, we are more in control of our environment, at least if we live alone. Living alone, we can just decide not to bring home foods we know we should not be eating. Living with others, we are presented with whatever unhealthy processed addictive foods someone else may have brought in. Or they may insist that we buy the sugared cereals, the desserts, the chips, and the snacks, and make life miserable if we don't.

At home watching television, we are again barraged with advertising carefully designed to make us want food. We may have just finished dinner, but the carefully tested commercials generate desire, and we get up and get a bowl of ice cream to watch with the show, even though we had pushed away from the table too full to eat another bite just a few minutes ago.

Another environmental variable is temperature. We tend to keep our buildings at a comfortable temperature and spend most of our time indoors. Exposure to cool temperatures (60°F) triggers brown and beige fat to burn fatty acids and glucose to keep us warm.

INFLAMMATION

Inflammation is a key part of the body's innate immune system. When cells are attacked by pathogens or irritants, or when cells are damaged, the body engages a protective response. Immune cells are mobilized, and the blood vessels in the affected area enlarge (vasodilation). The result is reddening, warming, and swelling, usually associated with some pain. The reddening and warming are what give inflammation its name, from the Latin for "burning." There are many names for problems involving inflammation, and we give the suffix -itis to them. Examples of inflammatory problems are arthritis, tonsillitis, urethritis, diverticulitis, vasculitis, laryngitis, and so on.

Acute inflammation is easy to spot. A bee sting, a mosquito bite, or a splinter causes pain, redness, and swelling in the area affected. Chronic (prolonged) inflammation can affect blood vessels and internal organs, and can be more difficult to notice. This chronic systemic inflammation is what is associated with obesity, diabetes, and metabolic syndrome. It is usually diagnosed based on blood tests for specific proteins associated with inflammation, such as C-reactive protein, interleukins, serum amyloids, and tumor necrosis factor alpha (TNF-alpha). Of particular interest to obesity are interactions with insulin, blood glucose, and the satiety hormone leptin.

Chronic systemic inflammation is what is associated with obesity and its related diseases (J. Kang et al. 2010). Inflammation can cause obesity, and obesity causes inflammation, in a deadly cycle. Eliminating either obesity or inflammation can help eliminate the other.

Obesity is a state of chronic inflammation (Dandona, Aljada, and Bandyopadhyay 2004). We can predict future obesity and diabetes risk by looking for inflammatory mediators (TNF-alpha and interleukin-6) in a blood sample. These molecules in the blood can interfere with insulin action, causing insulin resistance. This, in turn, reduces insulin's anti-inflammatory effects, resulting in even more inflammation. The inflammatory hormone TNF-alpha is produced by fat cells and impairs insulin signaling, causing insulin resistance (J. Kang et al. 2010). The more adipocytes (fat cells) we have, the more TNF-alpha we produce, and the more insulin resistant we become.

Fat cells, called adipocytes by doctors and scientists, are not just storage bags for energy. They are hardworking factories of over 50 different hormones (called adipokines) that affect the cells themselves, cells in the rest of the fat tissue (adipose tissue), and cells throughout the body when they are secreted into the bloodstream. The two most studied adipokines are leptin and adiponectin (Greenberg and Obin 2006).

Adiponectin increases sensitivity to insulin in muscles and the liver. It increases the burning of fatty acids in several tissues, such as muscle cells. This results in fewer fatty acids found in the blood. Adiponectin reduces the amount of glucose in the blood, again by burning it (compare this to insulin, which reduces glucose and fatty acids by storing them in fat cells). It burns glucose and fatty acids by activating the enzyme AMPK. This is the same enzyme activated by exercise, and this is why exercise increases insulin sensitivity and burns fatty acids.

You would think that as people get fatter, more adiponectin would be produced, since it is produced in fat cells. But actually, the opposite happens. As a rule, when we gain weight, we don't create new fat cells. We fill up the existing fat cells. They grow as they fill with triglycerides. As they distend, they become less effective at producing adiponectin.

As we get fatter, we produce less adiponectin. This causes us to become less sensitive to insulin and more prone to high blood sugar. These are the hallmarks of metabolic syndrome and type 2 diabetes. Low adiponectin also causes atherosclerosis through mechanisms involving adhesion molecules and the transformation of macrophages (immune cells) into foam cells that clog arteries. Many people who are trying to lose weight are doing so to prevent heart disease. Reducing body fat percentage reduces the inflammation that causes hardening of the arteries.

The other well-studied adipokine produced by fat cells is leptin. Unlike what happens with adiponectin, leptin levels grow along with the fat cells. It is, after all, a signal that the fat cells are full and that we should stop eating. It normally does stop us from feeling hungry. However, many of us eat for pleasure or eat mindlessly or out of boredom, long after we are no longer hungry. The fat cells fill up and produce even more leptin, which we continue to ignore. Eventually, the leptin levels are so high that the cells receiving the signal start to ignore it. We become insensitive to leptin. When we are insensitive to leptin, we are hungry all the time.

Leptin interacts with the immune system. People with protein energy malnutrition—people who are starving—have very low leptin levels. Low leptin levels lower the production of interferon-alpha, tumor necrosis factor alpha, and nitric oxide generated by immune cells called macrophages. These people are prone to infections, since their immune system is not functioning well (Dandona, Aljada, and Bandyopadhyay 2004). For the same reason, they are less prone to autoimmune diseases like colitis and arthritis. Women, with their higher body fat percentage, have more leptin than men have and are more prone to autoimmune disorders.

Waist circumference correlates with markers of inflammation. Fat cells expand when we overeat, and they can then start to leak glycerol and fatty acids. The macrophages then come in to clean up, and they release inflammatory chemicals (cytokines). The fat cells themselves help by releasing their own inflammatory cytokines. Chronic systemic inflammation is associated with atherosclerosis and insulin resistance. The former is a major cardiovascular disease, and the latter is a contributing factor in further increases in obesity by making us overeat even more.

Inside fat cells are molecules called perilipins. These surround the stored triglycerides, preventing enzymes called lipases from breaking the fats down into free fatty acids. However, as the fat cells swell up with extra fat, they can't keep up the production of perilipins. The result is the breakdown of fat into free fatty acids, and the release of those fatty acids into the bloodstream. Muscle cells detect these fatty acids and respond by lowering their sensitivity to insulin.

Macrophages live in fatty tissue, and their numbers increase as we get fatter, since one of their functions is to eat fat cells that are no longer functioning properly. The macrophages produce the inflammatory cytokines tumor necrosis factor alpha (TNF-alpha), interleukin-6 (IL-6), and several others, which not only cause inflammation but also reduce insulin sensitivity (Dandona, Aljada, and Bandyopadhyay 2004). The cells also start producing more coagulant proteins that cause blood clots. TNF-alpha is also expressed by the fat cells themselves, and the enlarged fat cells found in obese animals overexpress it (Kahn and Flier 2000).

TNF-alpha is part of a feedback loop against excess energy storage. It inhibits the storage of new fat in the adipocytes (fat cells) and it increases the breakdown of fat in those cells into glycerol and free fatty acids. It also makes tissues less sensitive to insulin (Kahn and Flier 2000).

Inflammation raises insulin levels. The body adjusts by becoming insulin resistant. But high levels of insulin cause leptin resistance, so an important satiety signal is ignored, and we feel hungry all the time. Of course, this causes more overeating. In a very real sense, once we become fat, it is hard to lose weight because the inflammation is making us continue to overeat. We're fat because we're fat. Worse, insulin resistance causes more inflammation, leading to another obesity cycle.

Blood tests can be done to check for C-reactive protein, an important marker for inflammation. Elevated levels indicate chronic systemic inflammation and increase the risk of heart attacks, strokes, high blood pressure, and muscle weakness.

The key inflammatory cytokine known as interleukin-6 is induced by hyperglycemia (high blood sugar). Hyperglycemia happens when sugar comes into the blood faster than insulin can respond and remove it. In healthy individuals, this happens when we get too much sugar or fast-digesting starch. The resulting inflammation causes insulin resistance, which means that we become even *more* susceptible to hyperglycemia. Another deadly cycle.

When we eat too much, for whatever reasons, we get fat. The fat produces inflammation. The inflammation promotes insulin resistance. Insulin resistance eliminates the anti-inflammatory effect of insulin, leading to more inflammation. Antidiabetes drugs such as thiazolidinedione (an insulin sensitizer) can reduce insulin insensitivity and thus reduce inflammation, reducing the risk of heart disease (Dandona, Aljada, and Bandyopadhyay 2004).

The saturated fat palmitic acid, the most common saturated fat found in plants and animals, is produced in the body when we get excess carbohydrates. High levels of palmitic acid in the blood lead to the production of the inflammatory cytokine interleukin-1 beta. Other saturated fats, such as medium-chain triglyceride (MCT) oil and coconut oil, have anti-inflammatory effects. Other anti-inflammatory fats are the omega-3 fatty acids EPA and DHA (found in fish oil), alpha-linolenic acid (found in walnuts, walnut oil, hemp oil, and flax oil), and the omega-6 fatty acid DGLA. Replacing high palmitic acid fats and high arachidonic acid fats with MCT oil and omega-3 fats can reduce inflammation. Replacing them with fats that are neither pro-inflammatory nor anti-inflammatory, such as the oleic acid in olive oil, can also help. Dairy fat contains trans-palmitoleic acid, which is anti-inflammatory and reduces blood lipid (triglycerides and cholesterol) levels and blood levels of C-reactive protein (Bernstein, Roizen, and Martinez 2014).

Other foods that promote inflammation are refined carbohydrates (sugar, white bread, pastries, etc.), fried foods (french fries), red meat, processed meat (hot dogs and sausage), margarine, shortening, and lard. Oxidized cholesterol in fried foods damages white blood cells, leading to inflammation.

Anti-inflammatory foods include tomatoes, olive oil, green leafy vegetables, nuts (almonds, walnuts), fatty fish (salmon, sardines), shiitake mushrooms, beets, and many fruits (blueberries, cherries, strawberries, oranges). Important anti-inflammatory spices include turmeric, cloves, rosemary, cinnamon, allspice, oregano, marjoram, sage, thyme, peppers, onions, garlic, and ginger. Green tea also has antioxidant and anti-inflammatory properties. Fermented foods such as pickles, olives, yogurt, and sauerkraut are anti-inflammatory. Dietary fiber is also anti-inflammatory.

FETAL PROGRAMMING

Fetal programming is a stress response where effects on the mother, such as stress caused by food insecurity, cause changes in the fetus that affect how the offspring react to food, reward, and stress. This kind of programming is not found only in the fetus. In studies of juvenile macaque monkeys whose mothers were exposed to food insecurity when the monkeys were three to five months old, researchers found increased BMI, waist circumference, and insulin resistance in the young (Bose, Oliván, and Laferrère 2009).

In studies of the 1944 Dutch famine survivors, low birth weight due to maternal malnutrition was associated with obesity and metabolic disease (such as diabetes) when the children grew to adulthood (Bose, Oliván, and Laferrère 2009).

Animal studies of fetal dietary restriction show increased levels of glucocorticoids in the adult animals. Prenatal exposure to these stress hormones leads to impaired glucose tolerance and insulin resistance in the adult animals. In humans, low birth weight is associated with smaller head circumference at birth, short stature, and excess cortisol in middle age. Low BMI at age seven is associated with higher cortisol levels and higher waist-to-hip ratios in adult men and women.

In studies of obesity in rats, mothers who were undernourished during pregnancy gave birth to pups that were significantly smaller at birth and that ate more than control rats did. The overeating increased as the rats aged and was accelerated by a high-fat diet. These rats also showed high insulin levels, high leptin levels, and high blood pressure (Vickers et al. 2000).

Fetal programming does not alter the genes themselves in the fetus. Instead, what is altered is the expression of the genes, in ways that adapt the fetus to changes in the environment during development (Breier et al. 2001). The key systems affected are the leptin and insulin systems, leading to insensitivities in both. Fetal programming is not just caused by famine. Obese mothers also give birth to children with appetite regulation problems and insulin resistance.

There is a J-shaped or U-shaped distribution of birth weight to adult obesity, with low birth weights and high birth weight both being associated with adult fat mass (McMillen, Adam, and Mühlhäusler 2005). There is also an association with the birth weight of both the mother and the father on the birth weight of the child. Adjusting the BMI of the mother before or during pregnancy reduces this association, indicating that it is fetal programming, not genetics, at work. Heavier mothers have heavier babies, and this causes obesity later in life (McMillen, Adam, and Mühlhäusler 2005).

Complications of pregnancies caused by type 2 diabetes, gestational diabetes, or simply impaired glucose tolerance are also associated with later obesity in the offspring. Studies of diabetic mothers show that half of them had children with weights in the 90th percentile at birth and at age eight (McMillen, Adam, and Mühlhäusler 2005).

Small babies tend to have lower BMI in adult life, but this is due to reduced muscle mass, not reduced fat. In fact, they have higher abdominal fat, and higher overall fat, than adults born at normal weight. As we have seen, abdominal fat is what puts people at risk of metabolic syndrome. These babies also have higher leptin levels later in life.

As the obesity epidemic increases, the number of obese mothers will increase, and thus the number of babies programmed to be obese in later life will increase. This is a positive feedback that is likely to have contributed to the current global obesity crisis.

LOW INCOME AND LOW EDUCATION

In the developing world, the populations with high social classes have greater rates of obesity (Mohamed et al. 2014). The opposite seems to be the case in the developed world. In the latter environment, the cheapest foods are those made from subsidized grains and sugar, and access to fresh whole foods in poor neighborhoods is more limited, so those with lower income often find it difficult to afford high-protein, high-fiber, fresh, and unprocessed foods. Those with less education many also tend toward obesity in the developed world, since many of them miss out on diet and lifestyle advice that counters the advertising from food manufacturers. It's difficult, though, to separate education from income as a factor, since they're so frequently linked.

However, many dietary interventions are both available in low-income areas and affordable. Dried beans are inexpensive, keep well, and are available in bulk. They are both high in fiber and high in protein, and low in quickly digested carbohydrate. The carbohydrate can be further countered by eating cooked bean dishes cold, so some of the starch becomes resistant to digestion, and by using vinegar dressings (in a cold bean salad) to reduce the digestibility of what starch remains. Frozen or canned vegetables are almost as good as fresh, in-season vegetables; they also keep well and are available year-round. (Fresh carrots are available year-round as well, and are often available in large quantities—for juicing—at very low prices, but these are often available only in well-to-do neighborhoods.) It is possible to eat a healthy diet on a strict budget. However, there are far too many people who run out of food money before the end of the month, no matter what foods they buy.

Low income and low social status are stressors in modern life, and, as we saw in the section on the stress response, the result is upper-body obesity, depression, and metabolic syndrome (Bose, Oliván, and Laferrère 2009). Low income is often associated with stressors such as job strain and sleep deprivation. The temptation to treat the stress response with comfort food is strong, as the alternatives such as exercise, meditation, yoga, and positive social interactions are less available to people with fewer means.

WHEN WE EAT AFFECTS HOMEOSTASIS

FASTING

Our bodies have a set point for how much fat we carry. If we reduce our food intake, the body reduces our metabolic rate in order to maintain the precious fat we have stored. If we have become insulin resistant, the high insulin levels put a priority on storing fat instead of burning it, and our set point rises.

So how do we reduce insulin resistance? We undo the cause of it. We became insulin resistant by having constant high levels of insulin. Each of the two components—the constant part, and the high levels part—can be fixed.

Whenever we eat we raise our insulin levels. Some foods raise it more than others, but *all* foods raise insulin.

The way to fix insulin resistance is to stop eating. Completely. Allow the body to have extended periods of very low insulin levels. This is how we reset the insulin sensitivity levels. We need to feel hungry. That means no food for a whole day (24 hours), or better yet, for 36 hours.

We already fast for 8 hours or more, every day. We don't eat a lot when we're sleeping. (If you do, that is a problem we don't cover in this book, but it sounds dangerous, and you should get some professional help.)

Since we start with a free 8 hours of fasting, all we need is another 16 hours. Just hold out until dinner. Or, once you have become accustomed to an occasional fast, stick it out until breakfast, giving you 32 hours of low insulin levels. Holding out until lunch gives you that 36 hours.

Humans, and pretty much all animals, have evolved to handle periods where food is not available. This is a normal state in the wild. Our ancestors didn't eat when they were ill, when winter stores ran out, during frequent famines, and often for religious reasons. We are built to handle fasting. It is only in the modern world, where food is everywhere, that we seem to think we have to eat all the time. Snacking is a modern invention of the food industry. Mothers used to tell their children not to eat between meals, lest it spoil their dinner. The reason we have any fat at all is to tide us over in lean times. Now is the time to use those fat stores.

We store energy in two forms: fat (triglycerides) and glycogen. Glycogen is a starch-like carbohydrate made up of many glucose molecules bound together. It is stored in the liver, where it can be quickly converted into glucose when it is broken down, in a process called glycogenolysis (a word that just means "breaking up glycogen"). Muscle cells also store some glycogen when prompted to by insulin.

After 24 hours without eating, glycogen stores in the liver are gone. This leaves the liver with only one way to make glucose, which is to use lactate, glycerol, or amino acids as a base. The amino acids come from proteins. The glycerol is the part of fat that isn't a fatty acid. Lactate is formed when glucose is used for energy without being burned first.

Making glucose from glycogen requires energy, meaning that the net energy received when the glucose is finally burned is lower than when free glucose from digestion of food is used.

It is obvious that if we stop eating, we will lose weight. The current record for fasting is 382 days (Stewart and Fleming 1973). A 27-year-old male who weighed 456 pounds underwent a medically monitored fast, consuming only noncaloric fluids and vitamins. Every day he submitted urine samples for analysis, and mineral supplements were added as indicated by the analysis. During this time, he felt well, had no symptoms of any illness, and walked around normally. He lost an average of 0.72

pounds per day. Several other long fasts have been recorded, from 210 days to 256 days, with similar results.

After a day or so of fasting, the liver is creating glucose from amino acids and glycerol. Fats are broken apart into glycerin and fatty acids. The glycerin is used to make glucose, and the fatty acids are used directly by muscles and organs. Some fatty acids are broken down into molecules called ketone bodies, which feed the brain.

After about 24 hours of not eating, we switch from burning stored glucose to burning fats. Blood glucose levels remain steady, since we are producing glucose from the glycerin in the fat. Insulin levels drop throughout the fast, and longer fasts allow the insulin to drop the most.

This drop in insulin is very important. Unlike dieting, which only prevents increases in insulin, fasting actually lowers insulin, which is key to getting back your insulin sensitivity.

After about three days of fasting, the body starts producing high levels of human growth hormone and adrenaline (Ho et al. 1988). The HGH maintains lean body mass, so your muscles aren't used as food. That only happens when the last of your fat stores are depleted, and we aren't going anywhere near that point. Growth hormone levels can double after five days of fasting.

The adrenaline maintains your metabolic rate, so we stay warm and have the energy to go out and find food. Adrenaline levels rise after 24 hours of fasting, and can rise by over 3 percent after 48 hours. This is not so much that you feel jittery and anxious, but it can increase your metabolic rate by 14 percent after four days. Note that this is the opposite of what happens when you simply decrease the amount of food you eat. You need to exhaust your stored glucose supplies to make this happen.

But we don't have to go anywhere near those extremes. We aren't racing. If you fast once a week, and lose three-quarters of a pound each time, that adds up to just under 40 pounds a year. Fast twice a week for 24 hours each, losing a pound a week, you drop 52 pounds in a year.

For many people, it is much easier to fast than it is to diet. It is easier to say no to all food than it is to try to limit meals to small portions.

The important thing to note is that when we switch from burning stored glucose to burning fat, that switch is accompanied by several hormonal changes that benefit weight loss. Just reducing calories, as in all the popular diet plans, does not give these benefits. Calorie reduction lowers the metabolic rate. Fasting does not, and extended fasting actually increases the metabolic rate. Calorie reduction does not correct insulin resistance. Fasting does.

The longer you fast, the more your appetite decreases. Your body becomes accustomed to the new regime and stops insisting that you eat. You will still feel hunger spurts, but less often.

When your stomach produces ghrelin, it does so in little spurts. It is a reminder, not a constant alarm. You feel a hunger spurt, but it goes away after a few minutes, and doesn't return for 20 minutes or more. It is easier to fast than most overfed people think it is. When you feel hungry, drink a glass of water. The hunger goes away.

For most obese people, after a couple days of fasting, hunger just vanishes altogether. The body has switched to burning fat, and it knows it is holding months of food.

Unlike calorie restriction dieting, there is an end to intermittent fasting. Instead of making a lifestyle adjustment for the rest of your life, you can just fast periodically. Most people find that this is much easier to do. It is easier to have a short period of "won't" power than a long period of willpower. Of course, some lifestyle changes, such as changing from a high-sugar, high-fat diet to a high-fiber, high-protein diet, or getting more exercise, have many benefits beyond weight loss and are highly recommended. But those changes are usually easier to maintain, since they increase well-being and don't cause a sense of sacrifice.

There are several other benefits to fasting than just losing weight and regaining insulin sensitivity. A 48-hour fast leads to a 50 percent reduction in reactive oxygen species generation by leukocytes in the blood (Dandona, Aljada, and Bandyopadhyay 2004). This reduces inflammation, which we have seen has large effects on obesity and diabetes. Nonobese test subjects given excessive levels of carbohydrate and fat had pro-oxidant and pro-inflammatory effects similar to what obese patients have as a normal condition.

During the first part of the fast, up to about five days, you lose water and salts. This is because high insulin levels cause the kidneys to store salt and water. As the insulin levels drop, you can lose almost two pounds a day, as you get rid of the bloating and water retention caused by high insulin levels. This also can lower blood pressure. Even short fasts of 24 hours will have this effect until insulin levels are under control.

Fasting creates a sense of well-being, almost a mild euphoria.

In the section on the hunger hormone ghrelin, we discussed many desirable effects of allowing the stomach to create the hormone:

- Ghrelin protects against atherosclerosis by acting on receptors in the blood vessels.
- Ghrelin is an antidepressant and reduces anxiety.
- Ghrelin inhibits insulin release after a meal.
- Ghrelin has anti-inflammatory effects and helps cells grow despite stress.
- Ghrelin is being considered as a therapy for colitis and sepsis.
- Ghrelin helps the stomach regenerate after injury to its mucosal layers.
- Ghrelin releases human growth hormone (HGH).
- Ghrelin modulates sugar and fat metabolism.
- Ghrelin regulates gastrointestinal motility (it makes food move through your digestive system better).
- Ghrelin protects neurons and cardiovascular cells.
- Ghrelin regulates the immune system.
- Ghrelin acts on the brain's hippocampus to increase learning and memory.
- Ghrelin helps you form new brain cells, even in adulthood.

All these benefits of ghrelin come with its other effect, making you feel hungry. Allow yourself to feel hungry between meals. You will be smarter, stronger, and healthier.

In the United States, about 60 percent of yearly weight gain happens during the holidays. Consider fasting once the leftovers are gone, to get back to your preholiday weight (or less). It can be very hard to diet when others are feasting. Go ahead and celebrate with them. If you allow yourself the indulgence, you can undo the harm when the guests have gone home. Let the skinny ones take home that extra pie and mashed potatoes.

INTERMITTENT CALORIE RESTRICTION

Just reducing daily calories by two-thirds for two days a week has been found to be as effective for weight loss as continuous dieting. In fact, intermittent dieting has been found to produce slightly better fasting insulin levels and slightly lower insulin resistance than continuous dieting (Harvie et al. 2010).

In a study of men fasting for 20 hours every other day, insulin sensitivity rose 700 percent (Halberg et al. 2005). Levels of adiponectin, the fat-burning hormone that increases insulin sensitivity, rose 37 percent. In this study, the subjects were told to eat abundantly on their nonfasting days so their body weight and body fat percentages would stay the same. Eating normally instead would cause weight loss and reduce fat.

We are able to make mice live longer by reducing their calorie intake dramatically. This is obviously very hard to do with humans in today's "food everywhere" environment. But one study in mice found that alternate-day fasting had similar effects, improving glucose management and brain cell resistance to stress, even though the mice maintained their body weight by eating more on nonfasting days (Anson et al. 2002). In another study of intermittent fasting, it was found to increase learning and memory (L. Li, Wang, and Zuo 2013).

Eating a single meal per day, even when the same number of daily calories is consumed, reduces fat more than consuming three meals a day. You get the benefits of fasting for 24 hours. Your insulin levels drop for that period, and you get your insulin sensitivity back.

Even when the same number of daily calories is consumed, eating more meals per day increases liver fat and abdominal fat more than eating three or fewer meals per day.

There are many approaches to intermittent fasting. Some of the popular ones are:

- Eating one meal a day. You still want to make it a healthy meal, but it can be 2,000 calories.
- Alternate between eating normally and eating only one meal every other day. Drink plenty of noncaloric liquids for two non-meals.
- Fasting completely every other day. Drink noncaloric liquids and perhaps a cup of vegetable broth.
- Fasting two days a week.
- Fasting one day per week. This can be for 24 or 36 hours or more.
- Eating only during one 8-hour period during the day (16 hours of fasting). This is basically just skipping breakfast every day. Lunch at noon, dinner before 8.

Eliminating snacks is important. Allow your insulin levels to drop between meals, no matter how many meals you end up eating. Make sure you are getting enough liquids and replace the salt you lose. A lack of fluid (dehydration) can make you feel weak or dizzy or give you headaches. It isn't the lack of food. Get enough magnesium, either from multivitamin pills or a soak in an Epsom salts bath.

When you aren't fasting, make sure you get plenty of fiber. Even when you are fasting, you can take pure fiber supplements: Metamucil or fiber capsules. Not only will this fill your stomach and quell hunger spikes, but it will prevent constipation.

Don't stuff yourself when you *do* eat. This can cause heartburn and acid reflux. Eat slowly so your body has time to notice how much food you have eaten and can signal you with satiety hormones. This generally takes about 20 minutes. Eating high-fiber foods that take longer to chew— salads, carrots, other vegetables—is a natural way to slow down a meal. Processed foods are designed to melt in the mouth and be eaten quickly, so you consume much more food before your body can tell you to stop. Eat more natural foods, not food that has been prechewed for you in a factory or in a blender.

When you do eat, eat healthy. Avoid sugars and processed foods. Fasting is not a license to eat anything you want on nonfasting days. We are trying to become more insulin sensitive. Eating sugar and wheat flour when we aren't fasting sabotages that. That's how we got here. Don't do that.

Drink plenty of noncaloric liquids and replace any missing salt and minerals.

Give yourself a month to get used to intermittent fasting. If it is difficult at first, stick with it, and it will get easier. Unlike dieting, studies have shown that intermittent fasting programs have an 80 percent compliance record (Hankey, Klukowska, and Lean 2015). People find it much easier to skip meals than to diet every day.

Pick carefully whom you tell about your fasting. Some people will try to sabotage your willpower. Others will be part of a beneficial support system. Find out which of your friends and coworkers are in each group, and control their access to information about your diet.

Stay busy. Work, hobbies, exercise, entertainment—all of these will take your mind off food during your fast. Get away from the kitchen, and away from tempting food sources and advertising. A walk in the park, a friendly game, even a drive in the car can take you away from temptations.

Don't binge when you aren't fasting. Just eat normally, or even a little more slowly than your old normal. Many people became overweight by not giving themselves enough time to become sated. Take it easy. Take the time to enjoy your meals.

Intermittent fasting works even better when combined with exercise.

A RELAXED DAY MAY HELP RAISE LEPTIN LEVELS

There is some support for the idea that getting an extra 1,000 calories per day one day per week can raise leptin levels. Higher leptin levels mean you don't feel hungry all the time. To make up for the extra calories, you can have a day with 1,000 fewer calories once a week. A day of fasting once a week more than makes up for this.

In popular diet literature, this is often called a cheat day, but the psychology of that term tends to make the practice fail. You aren't cheating. You don't have license to eat a bunch of sugar and fat. This is a period when you are getting more calories than you do during the rest of the week.

A cheat day is a day when you get to eat a cheat meal. You don't get to binge all day. Rather than relaxing the rules for an entire day, relaxing them for one particular meal per week can make a diet easier to follow. If you are entertaining, or being entertained, allowing yourself to relax the rules for that meal and enjoy the pasta, bread, or potatoes not only gives you something to look forward to during the week but can actually boost your metabolism and leptin levels as well.

You should find a diet where you can enjoy your meals, so you don't feel a need to "cheat." A high-protein, low glycemic index diet is satisfying to the 67 percent of people who have the genes that make that diet successful. For the other third of humanity, there are other diets that also work, as well as interventions such as adding vinegar when you eat starches, avoiding sugar, fasting, avoiding between-meal snacks, and so on. All these diets and interventions allow you to eat foods you enjoy, so you can feel comfortable and not feel the need to binge.

In the scientific literature, there is support for a mild form of fasting called time-restricted feeding, or TRF, a form of intermittent fasting (discussed in more detail in the preceding section, page 180). This is where you eat only during a particular time window each day. This is typically an eight- or nine-hour period, but the longer the fasting period, and thus the shorter the feeding period, the more weight loss occurs. In this regimen, adding a weekend of eating whenever you like has been shown to retain most of the benefits of a weeklong routine, at least in mice (Chaix et al. 2014).

A similar protocol is the 90/10 rule. Allowing yourself a relaxed meal one time out of 10 allows you to have something to look forward to and makes the diet less forbidding and onerous. The limit of one in 10 helps to prevent bingeing and blowing the plan.

Whether cheat meals are right for you may depend on your genes (Poehlman et al. 1986). Overfeeding affects people in different ways, depending on differences in their genetic makeup. Monitor your results, and if the reward meal isn't working for you, work on making every healthy meal rewarding instead.

Many of the studies done on overfeeding have been done on lean animals—usually rats or mice—or on lean people (Dirlewanger et al. 2000).

Overweight and obese people react differently to challenges to their insulin and glucose levels. People who are insulin resistant will not react favorably to a sudden glucose challenge, while lean individuals react with increased energy expenditure and increased leptin levels. Even lean people don't benefit from overfeeding on fats. Only carbohydrate overfeeding provided any metabolic and leptin boost.

If you have a diet that is working for you, you should definitely not add any cheat meals. Don't fix it if it ain't broken. But if you are finding yourself unable to stick with a particularly restrictive regime, adding an incentive reward just might help.

HOW TO TELL WHICH OF THE WEIGHT-GAIN FACTORS APPLY TO YOU

Some questions to ask yourself:

- Am I addicted to sugar?
- Do I eat a lot of processed foods and foods made from wheat flour?
- Do I get enough sleep?
- Am I under a lot of stress?
- Do I get enough exercise?
- Do I keep mindlessly eating as long as there is food in front of me?
- Do I allow three to four hours for insulin levels to drop between meals?
- Do I eat so quickly that the satiety signals don't have time to tell me to stop eating?
- Am I impulsive?
- Am I conscientious?
- Do I get too much blue light from TV or computer screens after dark?
- Do I have obesity-related genes?

- Do I have the right gut bacteria?
- Do I eat enough fiber?
- Do I drink too much alcohol?
- Do I eat for pleasure when I'm not hungry?
- Do I have a low metabolic rate?
- Am I depressed or on antidepressants?
- Do I have chronic inflammation?
- Am I constantly under a lot of stress?

GENETIC TESTING

There are many companies that will test your DNA and give you results you can screen for obesity-related genes. One of the more well-known companies is 23andMe. Once you get the results, you can browse through the data on a web page. An example of one result, for the rs1421085 location on the FTO gene, is shown below:

Genes	Marker (SNP)	Genomic Position	Variants	Your Genotype
FTO	rs1421085*	53800954	C or T	C / C

In this case, the person was homozygous (both father and mother contributed) for the variant where a cytosine replaced a thymine (a C replaced a T), which makes this gene unable to function. The result is that cells that might have turned into beige fat cells that burn fat instead turn into white fat cells that store fat.

The test results will often also include a huge list of the SNPs (single nucleotide polymorphisms—changes in the gene where a single nucleotide [an A, C, G, or T] is changed), and which variant you have. A small portion of such a list is shown on the next page.

This shows the SNP rs987237, where the AA variant means you will do well (lose weight) on a low-carbohydrate, high-protein diet. The AG and GG variants do not benefit as much from this diet. Not surprisingly, the AG and GG variants are associated with obesity.

Other genome sequencing companies besides 23andMe are FitnessGenes, Genos, Health

	6	50715...	
...40	6	50719347	
...495	6	50727940	
...83818	6	50756951	
...7772880	6	50762920	
...s2206272	6	50765352	G...
rs9367415	6	50782047	AC
rs2143081	6	50782834	AG
rs2206271	6	50786008	TT
rs6930924	6	50790633	GG
rs13216733	6	50791482	--
rs987237	6	50803050	AA
rs2817394	6	50803944	TT
i5001019	6	50805764	CC
rs2635727	6	50820940	CT
...s4715209	6	50829511	C...
...9381908	6	50836750	...
...1970011	6	50845787	
...4001	6	50850065	
...4	6	50850592	
	6	508570...	

Nucleus, Map My Gene, Nutrigenomics, GenePlanet, Gonidio, DNAFit, New Life Genetics, Genetic Health, LifeGenetics, Labrix, Pathway Genomics, IB Biotech, and Futura Genetics—the list grows almost daily.

BLOOD TESTS

Blood tests are available that can lend insight into the causes, and thus possible treatments, of obesity. Some of the more common ones are:

- Hemoglobin A1c test
- Fasting glucose test
- Oral glucose tolerance test
- Fasting lipid panel
- Alanine aminotransferase (ALT) test
- C-reactive protein test
- Thyroid stimulating hormone (TSH) test

The hemoglobin A1c test is usually a test for diabetes, as it tells you what your average glucose levels were over the previous three months. Blood glucose reacts with many proteins, and hemoglobin is a protein. The result is glycated hemoglobin, which this test looks for. The image on the next page shows the structure of a glycated hemoglobin protein.

You can see the iron-centered four-ring structures inside, and a few tiny glucose molecules, such as the one in the center of the middle cluster.

Red blood cells live only about four months, so the average level of glycated hemoglobin gives an estimate of the average blood sugar levels over the previous three months or so, depending on the current rate of blood cell replacement. Blood cells are replaced faster if we need to replace blood after losing it to accident, surgery, or blood donation, so the test will give false results in these situations.

The test results are also more sensitive to recent blood glucose levels than to older ones, so it is more accurate if the diet or lifestyle has not changed in the last three to four months.

The hemoglobin A1c test can be used as a hint to the levels of advanced glycation end products, since they are produced by the same processes.

The fasting glucose test measures sugar in the blood after a nighttime fast. If blood sugar levels are high after fasting, this indicates a problem in glucose metabolism.

The oral glucose tolerance test starts with a fasting glucose test, then you drink a sugary solution, and your blood sugar levels are tested over the next two hours. High blood sugar levels after two hours are an indication of diabetes and poor glucose metabolism.

The fasting lipid panel tells you about your blood cholesterol and triglyceride levels. It is used to determine the risk of cardiovascular disease. It is done by taking blood after a nighttime fast, so that it won't reflect the most recent meal as much, but what you are actually interested in is the average levels of fats in the blood throughout a typical day. The numbers given by the test are for total cholesterol, high-density lipoproteins (HDL), low-density lipoproteins (LDL), very low density lipoproteins (VLDL), all non-HDL cholesterol, the ratio of total cholesterol to HDL, and the total amount of fat in the blood.

Fat does not mix with water, and blood is mostly water. For fats to be transported in the blood, they must first be broken down into tiny droplets, and these droplets must be surrounded by a membrane that keeps them from clumping back together. The membrane is made up of proteins, cholesterol, and a molecule called a phospholipid, a kind of detergent, with a water-loving phosphorus atom at one end and a fat-loving oily tail at the other. This detergent surrounds the fat droplet, with the oily ends in the fat and the water-loving ends facing the water. The proteins help keep the whole thing together and interact with receptors on cells in the liver and other tissues. The cholesterol also has a water-loving end and a fatty end, and it fits in with the phospholipids and helps keep the membrane flexible and strong.

When we eat fats, the small intestine breaks them up and wraps them up in cholesterol-phospholipid-protein packets called chylomicrons. These are large fat droplets, and they get transferred in the blood until they reach the liver. The liver repackages them as smaller VLDL droplets, which are carried to the fat cells for storage or to other tissues for burning. As the fats in the droplets are gradually depleted by cells that cleave off fatty acids, the droplets become denser, since fats are less dense than the membrane around them. These lipoprotein droplets make their way back to the liver, where they are reused to make new VLDL droplets. Some of the LDL particles don't make it to the liver but are eaten by immune cells called macrophages. These can turn into what are called foam cells and stick to artery walls, causing atherosclerosis. This is why we call LDL the "bad" cholesterol.

The liver also makes HDL droplets that have very little fat in them. Their job is to pick up cholesterol from tissues in the body and transport it to other tissues, where the cholesterol is needed to build cell walls. This function is why we call HDL the "good" cholesterol.

When we eat foods containing cholesterol, the liver doesn't need to produce as much cholesterol itself. The total amount of cholesterol in the body doesn't change much due to dietary cholesterol levels. This is why we no longer worry about eating foods like eggs, which have more cholesterol than, say, carrots. Every cell in our body needs cholesterol to make its cell walls, and we will make cholesterol if we don't get enough in the diet.

The alanine aminotransferase (ALT) test detects problems with the liver, such as cirrhosis or hepatitis, caused by alcohol, drugs, or viruses. In concert with the other blood tests, it can be used to diagnose nonalcoholic fatty liver disease, whereupon an imaging test such as ultrasound, MRI, or CAT scan is used to confirm it.

The C-reactive protein test is another test to determine risk of heart disease. The liver produces a molecule called C-reactive protein in response to inflammation. High levels of this molecule in the blood tell you and your doctor that you have inflammation from some source. It does not tell you if that source is a bee sting or atherosclerosis in coronary arteries. This test is usually more predictive of heart problems than the lipid panel (cholesterol test), especially for women.

Since one of the important causes of obesity is inflammation, this test can help you decide a course of action. Of course, obesity is also a cause of inflammation, so you might want to consider anti-inflammatory diet suggestions even without bothering to get the test. Increasing omega-3 fats, decreasing inflammatory omega-6 fats, and getting lots of anti-inflammatory colorful plant molecules (green, orange, or red veggies; turmeric; capsaicin; etc.) is never harmful.

The thyroid stimulating hormone (TSH) test tells you if you have a healthy thyroid gland. An underactive thyroid can cause weight gain, fatigue, dry skin, feeling too cold, or constipation. An overactive thyroid causes weight loss, rapid heart rate, nervousness, diarrhea, and feeling too hot. While it is unlikely that your weight problem is due to an underactive thyroid, this is an easy condition to test for and to treat.

STOOL ANALYSIS

Testing the stool can tell us whether we have a low diversity of intestinal microbes (bad), and what the ratio of Firmicutes to Bacteroidetes is. It can also detect harmful organisms as well as check for colon cancer and parasites. It can detect disease of the digestive tract, liver, or pancreas. Larger samples collected over 72 hours can be used to check for undigested fat and protein, indicating poor absorption in the digestive tract.

There are several companies that will do a microbial ecology test. Below is part of one page of such a test, showing relative amounts of various intestinal microbes:

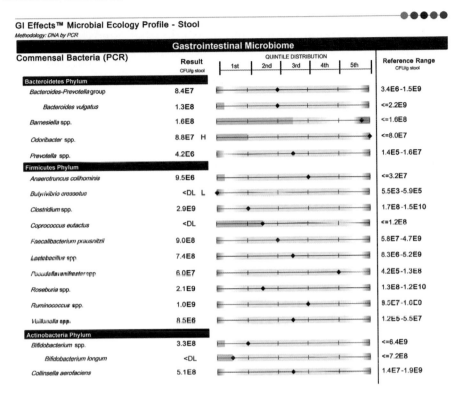

A crowd-funded program called the American Gut Project (http://americangut.org/) will sell you a stool sample kit; after the sample is submitted and analyzed, the results are posted on a website you can log into. Your sample results are added to a growing database of microbial data that can be studied by scientists to learn more about our symbiotic little friends.

THE POWER OF FOOD

The Power of Food Scale helps us to determine the psychological impact of living in a food-abundant environment (Cappelleri et al. 2009). The questionnaire is shown on the next page.

1. I find myself thinking about food even when I am not physically hungry
2. When I'm in a situation where delicious foods are present but I have to wait to eat them, it is very difficult for me to wait
3. I get more pleasure from eating than I do from almost anything else
4. I feel that food is to me like liquor is to an alcoholic
5. If I see or smell a food I like, I get a powerful urge to have some
6. When I'm around fattening food I love, it's hard to stop myself from at least tasting it
7. I often think about what foods I might eat later in the day
8. It's scary to think of the power that food has over me
9. When I taste a favorite food, I feel intense pleasure
10. When I know a delicious food is available, I can't help myself from thinking about having some
11. I love the taste of certain foods so much that I can't avoid eating them even if they're bad for me
12. When I see delicious foods in advertisements or commercials, it makes me want to eat
13. I feel like food controls me rather than the other way around
14. Just before I taste a favorite food, I feel intense anticipation
15. When I eat delicious food, I focus a lot on how good it tastes
16. Sometimes, when I'm doing everyday activities, I get an urge to eat "out of the blue" (for no apparent reason)
17. I think I enjoy eating, a lot more than most other people
18. Hearing someone describe a great meal makes me really want to have something to eat
19. It seems like I have food on my mind a lot
20. It's very important to me that the foods I eat are as delicious as possible
21. Before I eat a favorite food, my mouth tends to flood with saliva

THE THREE-FACTOR EATING QUESTIONNAIRE

This is a questionnaire designed to determine scores for three categories of psychological or emotional components of eating behavior (FLVS Study Group 2004). Those three factors are *uncontrolled eating, cognitive restraint*, and *emotional eating*. The questionnaire is shown here:

1. When I smell a sizzling steak or juicy piece of meat, I find it very difficult to keep from eating, even if I have just finished a meal.
    Definitely true (4) • Mostly true (3) • Mostly false (2) • Definitely false (1)

2. I deliberately take small helpings as a means of controlling my weight.
    Definitely true (4) • Mostly true (3) • Mostly false (2) • Definitely false (1)

3. When I feel anxious, I find myself eating.
    Definitely true (4) • Mostly true (3) • Mostly false (2) • Definitely false (1)

4. Sometimes when I start eating, I just can't seem to stop.
    Definitely true (4) • Mostly true (3) • Mostly false (2) • Definitely false (1)

5. Being with someone who is eating often makes me hungry enough to eat also.
    Definitely true (4) • Mostly true (3) • Mostly false (2) • Definitely false (1)

6. When I feel blue, I often overeat.
    Definitely true (4) • Mostly true (3) • Mostly false (2) • Definitely false (1)

7. When I see a real delicacy, I often get so hungry that I have to eat right away.
    Definitely true (4) • Mostly true (3) • Mostly false (2) • Definitely false (1)

8. I get so hungry that my stomach often seems like a bottomless pit.
    Definitely true (4) • Mostly true (3) • Mostly false (2) • Definitely false (1)

9. I am always hungry so it is hard for me to stop eating before I finish the food on my plate.
    Definitely true (4) • Mostly true (3) • Mostly false (2) • Definitely false (1)

10. When I feel lonely, I console myself by eating.
    Definitely true (4) • Mostly true (3) • Mostly false (2) • Definitely false (1)

11. I consciously hold back at meals in order not to gain weight.
    Definitely true (4) • Mostly true (3) • Mostly false (2) • Definitely false (1)

12. I do not eat some foods because they make me fat.
    Definitely true (4) • Mostly true (3) • Mostly false (2) • Definitely false (1)

13. I am always hungry enough to eat at any time.
    Definitely true (4) • Mostly true (3) • Mostly false (2) • Definitely false (1)

14. How often do you feel hungry?
    Only at meal times (1) • Sometimes between meals (2) • Often between meals (3) • Almost always (4)

15. How frequently do you avoid "stocking up" on tempting foods?
    Almost never (1) • Seldom (2) • Usually (3) • Almost always (4)

16. How likely are you to consciously eat less than you want?
    Unlikely (1) • Slightly likely (2) • Moderately likely (3) • Very likely (4)

17. Do you go on eating binges though you are not hungry?
    Never (1) • Rarely (2) • Sometimes (3) • At least once a week (4)

18. On a scale of 1 to 8, where 1 means no restraint in eating (eating whatever you want, whenever you want it) and 8 means total restraint (constantly limiting food intake and never "giving in"), what number would you give yourself?
    1 or 2 on the scale (1) • 3 or 4 on the scale (2) • 5 or 6 on the scale (3) • 7 or 8 on the scale (4)

The *cognitive restraint* scale is composed of items 2, 11, 12, 15, 16, and 18. The *uncontrolled eating* scale is composed of items 1, 4, 5, 7, 8, 9, 13, 14, and 17. The *emotional eating* scale is composed of items 3, 6, and 10.

You can answer the questionnaire and score it yourself (take the average of the answers in each of the three categories). That will give you some idea of whether you have problems in any category.

Knowing that you have a problem with emotional eating, for example, you might decide to fix the root problem (talk to a friend other than Ben & Jerry if you're lonely) instead of trying to eat your way out of it.

If you have an issue with restraint, you might try interventions such as weighing or photographing your food before you eat, using smaller plates, always eating with someone who can help monitor portions, and so on.

If you score high in uncontrolled eating, you might look for tricks to help you stay in control. Fasting is one method—it is much easier to say no to *everything* for a day than to try to measure and control each bite. Fasting also helps to fix underlying metabolic problems that might be causing the uncontrolled eating (not everything is about psychology; sometimes biology is to blame).

MEASURING WAIST-TO-HIP RATIO

The hip circumference is measured by using a flexible tape measure across the buttocks where they extend the most in the back. The waist circumference is measured by placing the tape around the waist at the midpoint between the bones of the ribs and the bones of the pelvis—this is about an inch above the navel, but in obese people, this may not be the narrowest place. The tape should always be parallel to the floor. The tape should not stretch, and the tension should be light (100 grams). Measure after letting out a breath.

Divide the waist circumference by the hip circumference to get the waist to hip ratio. For example, if your waist measurement is 38 and your hip measurement is 42, your waist-to-hip ratio (WHR) is 0.9.

Normal values for men are less than 0.9. For women, normal values are less than 0.8. Overweight for men is 0.9 to 0.99, and 0.8 to 0.84 for women. Obese for men is over 1.0 and over 0.85 for women.

ACTIVITY MONITORS

There are a large number of these devices on the market. Worn on the wrist, they can measure the number of steps you have taken each day so you can be motivated to get more exercise, or monitor movement at night, to tell you how many hours of sleep you are getting.

Many of these devices sync with your cell phone and can upload data to the Internet, where you can examine it on a large screen or share it with weight loss or exercise partners so you can motivate one another through friendly competition.

Some have GPS, so that in addition to footsteps, they can tell how far you cycled on a bike or how far up the hill you have climbed.

It is important not to lose focus on your behavior, however. Fitness trackers may allow some people to ignore how much they eat or how much they actually exercise, compared to people who track their exercise by hand, say, by entering information on a website. The extra effort involved in keeping an activity log by hand can help focus on the goals, and on the activities needed to reach them. Competing with friends who are also trying to get more exercise may help here, especially if the friends are close to your own performance levels, or you have a large enough number of them that you aren't always at the bottom of the list.

BODY COMPOSITION

As with activity monitors, there are a number of companies offering bathroom scales that measure not just weight, but body fat percentage, water percentage, protein percentage, heart rate, and arterial blood flow, and calculate your BMI. They can calculate how much you have gained or lost since you last weighed yourself, or since the same time the previous day. The better ones also send the data to your phone or to a website, so you can see long-term trends and performance graphs.

Below and on the next page are some typical graphs from the webpage of one Internet-connected bathroom scale:

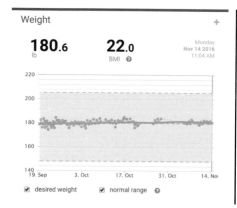
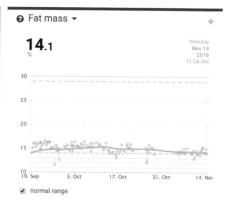

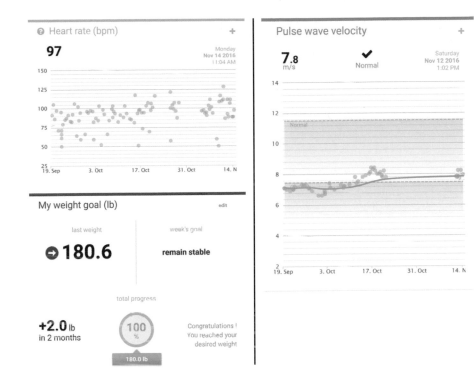

SIGNS OF INSULIN RESISTANCE

Insulin resistance is usually tested by a doctor, but there are signs you can look for yourself. You may have taken a cholesterol test at a drugstore or purchased a blood glucose test kit there, along with your blood pressure monitor. The signs to look for are:

- Skin tags
- Loud snoring
- Excessive daytime sleepiness
- Sleep apnea (waking up to take a breath)
- Ovarian cysts (polycystic ovary syndrome, or PCOS)
- High blood pressure (hypertension)
- High blood cholesterol
- High blood sugar
- High blood triglycerides
- Acanthosis nigricans (dark skin in body folds and creases)

REMEDIES

DIET REMEDIES

Reduce or Eliminate Sugar and Flour

Of course, the first dietary intervention should be to eliminate sucrose and fructose from the diet. No sugar. No agave nectar (made almost entirely out of fructose, and thus much worse than plain sugar).

Next would be eliminating wheat flour and products made from it. Wheat flour is mostly amylopectin A, the form of starch that is most rapidly digested into glucose. It is also ground into an extremely fine powder, which allows it to be digested very quickly, since the surface-to-volume ratio is so high.

Fiber

Next is increasing the amount of fiber in the diet. Green leafy vegetables, fruits, legumes (beans, peas, lentils), cruciferous vegetables such as broccoli and cauliflower, carrots, artichokes, and nuts are excellent sources. In general, the darker the produce, the more fiber is in it. Lower down on the list are sweet potatoes, and then russet and red potatoes. These are better if they are eaten cold, after a night in the refrigerator to allow the amylose starch to crystallize and become resistant starch, which acts like fiber.

You can add oat bran to recipes to increase the fiber content, as well as konjac powder, psyllium powder, or other pure fiber additives.

When eating fiber, remember to get lots of water with it. Much of the benefit of fiber is that it retains water and stays in the stomach longer, satisfying hunger and trickling the food through the digestive tract so blood sugar doesn't spike.

Water

Staying hydrated (getting lots of water) helps in several ways. It fills the stomach, relieving hunger. But since humans have a difficult time determining whether they are hungry or thirsty, they will often eat when they should be drinking. Drinking water instead of eating something when you think you are hungry is a good way to avoid getting calories when you aren't hungry. Eating when you aren't hungry is a common route to obesity.

These three diet interventions—eliminate sugar and flour, greatly increase fiber, get enough water—are the most important. The interventions that follow, while they each have measurable effects, are much less efficient at helping you lose weight. Don't count on magic ingredients to make you thin. That said, this book is about all the many contributing factors, and each little bit can help.

Calcium

There is an inverse association between dietary calcium intake and obesity. People who are deficient in calcium are more likely to be obese (Astrup 2008). Test subjects taking a gram of calcium per day were 11 percent less likely to gain weight than subjects taking placebos. Dairy calcium is even better than that in pills. It binds to fat, so some of the ingested fat is not digested and gets flushed down the toilet. Levels of fat in the blood after a meal are lower if there is sufficient calcium.

A lack of calcium makes people hungry for calcium-rich foods. This makes it harder to stick to a reduced-calorie diet.

Curcumin

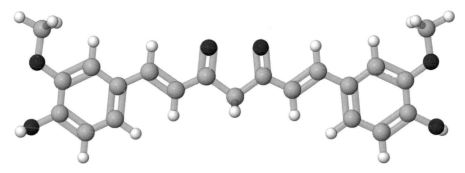

Curcumin, known to chemists as diferuloylmethane, is the yellow dye in the spice turmeric. Curcumin reduces obesity-related inflammation by preventing the fat cells from producing inflammatory molecules (J. Kang et al. 2010). It is those inflammatory molecules that cause insulin resistance. This is how curcumin helps in the treatment of diabetes, apart from its role in reducing fat.

A problem with research involving curcumin has recently come to light (Baker 2017). Many of the tests done to detect whether a drug or other molecule has a biological effect involve using fluorescent assays. These tests detect the light emitted by fluorescent molecules used to track the chemistry going on. But curcumin itself is fluorescent, and this can cause false positive indications from the test. So the studies in the following paragraphs need my cautionary language, since they have not yet been evaluated to see if they are affected by this problem. Where there is no cautionary language, the effect is known to be real.

Curcumin may interact with certain specific proteins in fat cells, pancreatic cells, muscle cells, liver cells, and the immune cells called macrophages. In those cells, it may suppress the protein transcription factor nuclear factor-kappa B, which controls the transcription of DNA and the production of immune cell signaling proteins (cytokines), and determines cell survival. It may also suppress STAT-3 (signal transducers and activator of transcription 3), the protein that allows cells to respond to inflammatory cytokines like the interferons and interleukin-5 and interleukin-6. Curcumin could suppress the Wnt/beta-catenin protein that regulates step cells and determines how and whether they become specialized cells such as fat cells (Shehzad et al. 2011).

Curcumin's effects are not limited to proteins. It may activate the peroxisome proliferator-activated receptor gamma (PPAR-gamma), a receptor on the surface of fat cells that controls their proliferation, their uptake of fats from the blood, and their production of new fat. This activation helps to clear the blood of fats and glucose, reducing high blood sugar levels and high cholesterol levels.

Curcumin could also activate nuclear factor-like 2 (Nrf2), a transcription factor that regulates the production of antioxidant molecules and protects against oxidative damage to cells caused by injury or inflammation.

Curcumin reduces levels of leptin and the inflammatory adipose-derived hormone resistin, both of which are produced by fat cells (Shehzad et al. 2011). Resistin increases levels of the bad form of cholesterol (LDL), and degrades liver receptors of LDL, making it harder for the liver to clear the blood of bad cholesterol. Resistin accelerates the formation of atherosclerosis, contributing to heart disease.

Curcumin increases levels of adiponectin, the hormone that slows glucose production, burns fat, decreases inflammation, and increases sensitivity to insulin.

Curcumin is thus a safe and inexpensive way to help control obesity and many of the obesity-related metabolic diseases. Including lots of curcumin or turmeric in the diet is a cheap way to protect yourself from many of these problems.

Capsaicin

In people with functioning beige fat, capsaicin and capsinoids have been found to activate brown and beige fat to burn calories for heat (Saito and Yoneshiro 2013). Nerves connecting the gut to the brain (the vagus nerve) control whether fat is deposited under the skin or centrally around the organs (abdominal fat). Capsaicin stops the stimulation of the vagus nerve caused by cholecystokinin and enhances expression of adiponectin and its receptor (Leung 2014).

Consumption of foods containing capsaicin is associated with reduced occurrence of obesity. In rural Thailand, people eat a diet that contains 0.014 percent capsaicin. When rats were fed that much, they did not eat less, but their visceral fat dropped by as much as 29 percent (Leung 2008). This occurs because the vagus nerve increases blood flow to the intestines and decreases blood flow to visceral fat deposits. This speeds digestion but inhibits fat formation.

Capsaicin also improves the rate at which fat is broken down (lipolysis) after exercise, and it increases satiety after a meal, while decreasing fat intake. After taking a capsule containing 2.56 milligrams of capsaicin per meal (1.03 grams of red chili pepper, 39,050 Scoville heat units), subjects had no diet-induced thermogenesis drop; they burned the same energy, despite getting 25 percent fewer calories while dieting (Janssens et al. 2013). These subjects also burned more fat.

Capsaicin inhibits macrophage responses to the signal from fat cells, reducing obesity-related inflammation (J. Kang et al. 2010). It also reduces inflammation in the liver, preventing fatty liver disease.

You may not even need to *eat* capsaicin to get the benefits. Topical application of a capsaicin cream to mice limited fat accumulation, reduced inflammation, and increased insulin sensitivity (G. Lee et al. 2013). Such creams are sold over the counter for relief of arthritis pain.

Sweet peppers (*Capsicum annum*) have a similar molecule, called capsiate, which has the same effects as capsaicin but without the hot pungency.

Capsaicin is also good for your heart. It stimulates the blood vessels to release nitric oxide, which makes them widen, lowering blood pressure (McCarty, DiNicolantonio, and O'Keefe 2015). It also has protective antioxidant effects in blood vessels and the liver and reduces atherosclerosis, metabolic syndrome, diabetes, fatty liver disease, cardiac hypertrophy, and stroke risk.

Black pepper and ginger have different but similar molecules, and have some, but not all, of the effects of capsaicin.

Garlic and Onions

Garlic contains biologically active compounds that affect both fat tissue and the tissues in the body that burn fat for heat, such as brown fat, and cells in the liver, muscles, and white fat (M. Lee et al. 2011). Garlic also affects genes associated with the storage of fat in fat cells. One study looked at supplementing the diet of mice with 2 or 5 percent garlic (as a percentage of the total weight of food). For a human, a 600-gram meal would thus have to have 12 to 30 grams (one ounce) of garlic. For every meal.

For many people that sounds like heaven, but it might not for those in the carpool the next morning. Fortunately for them, onions have many of the same compounds in them, and a 5 percent onion diet is not that different from meals we are used to eating. A study of diabetic obese rats found that a 3 to 5 percent onion diet led to lower blood glucose levels, lower insulin resistance, and lower fats and cholesterol in the blood (Yoshinari, Shiojima, and Igarashi 2012).

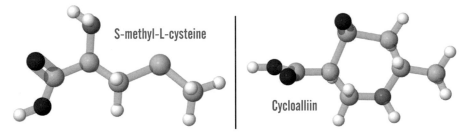

S-methyl-L-cysteine

Cycloalliin

That study used two specific sulfur-containing molecules found in onions, S-methyl-L-cysteine and cycloalliin, shown above, along with six similar sulfur-containing molecules, all found in onions. The rats eating the onion extract lost weight, and in particular lost weight in the liver, where obese rats store fat, as humans with nonalcoholic fatty liver disease do. The diabetic rats improved their glucose tolerance, apparently due to reduced levels of fatty acids in the blood and muscles, and thus lowered their insulin resistance. The effects were dose dependent, with greater effects seen at higher dosages.

Another study looked into non-sulfur-containing molecules found in onion peel tea. Most of us throw away the peel of the onion, but it contains catechins, biologically active compounds that have many health benefits. One such catechin is the flavonoid quercetin, shown below.

Like the sulfur-containing compounds, quercetin regulates fat metabolism and suppresses high blood sugar levels. Flavonoids and polyphenols are found in many plant foods and have many health benefits, such 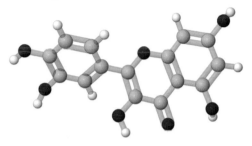 as antioxidant effects, anti-inflammatory effects, and fat-regulating effects. Onions are one of the foods in our diet that are richest in flavonoids.

The rats in the study were given 5 percent of their diet in onion peel tea. This amounted to between 5 and 6 milligrams per kilogram of body weight, which for a 180-pound human would mean less than half a gram (408 milligrams) per day. The rats lost weight and fat mass and had lower levels of leptin as a result.

Green Tea and Grape Seed

Phenolic compounds such as epigallocatechin gallate (EGCG), resveratrol, quercetin, and kaempferol inhibit pancreatic lipase (Sergent et al. 2012). Rutin and o-coumaric acid were found to be effective at reducing obesity in rats fed a high-fat diet (Hsu et al. 2008) in doses of 100 milligrams per kilogram of body weight. For a 180-pound man, that would come to 8.2 grams per day.

EGCG and other catechins have been shown to reduce fat cell proliferation and fat production; reduce fat mass, body weight, and fat absorption in the gut; lower blood triglycerides, free fatty acids, and cholesterol; lower blood glucose, insulin, and leptin levels; and increase thermogenesis, the burning of fat to provide body heat (Wolfram, Wang, and

Thielecke 2006). That's quite a lot of benefit from some fairly simple molecules. The structure of EGCG is shown to the right.

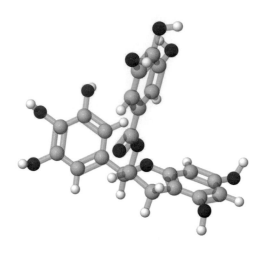

You can simply drink green tea, or you can buy it in concentrated form in capsules as a supplement, often sold as "fat burner" pills. These contain lots of EGCG, along with a coffee cup's worth of caffeine. Together the two molecules reduce hunger, as well as having all the above effects, and this makes fasting easier, should you be trying that as well.

Grape seed and skin extract, as well as grape seed oil, contain catechins. These have been found to have antiobesity and several cardioprotective effects, reducing inflammation, enlarged heart, and levels of C-reactive protein, a marker of inflammation (Charradi et al. 2011). The oil of Muscadine grape seeds was studied for its benefits in reducing fat production and inflammation (L. Zhao et al. 2015). It is a source of a type of vitamin E called tocotrienols. It has been found to reduce the expression of genes and proteins required to make fat.

Berberine

Berberine is another yellow plant dye, like curcumin. Like caffeine, ephedrine, morphine, cocaine, nicotine, strychnine, mescaline, and dopamine, it is

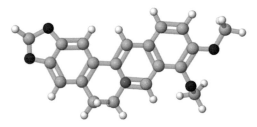

an alkaloid (a ring molecule containing a nitrogen atom, shown in blue in the structural formula shown above).

Berberine is found in many different plants. It gives the yellow color to tree turmeric (*Berberis aristata*) bark and roots, and is found in other barberry species (e.g., *Berberis vulgaris*), as well as California poppies, Oregon grape, goldenseal, and yellowroot. It has a bitter taste, like most alkaloids.

Berberine is antibacterial, antifungal, and antiviral, and is used as a topical anti-infective. It is rapidly digested, making it less effective for these purposes when eaten. However, its metabolites (breakdown products) are biologically active and are the components that give it anti-inflammatory, antiobesity, and antidiabetes properties. It also lowers blood pressure and acts as an anticonvulsant and a mild sedative. It lowers blood triglycerides and total cholesterol and increases insulin sensitivity by 50 percent (J. Yin, Xing, and Ye 2008). Like most brightly colored plant extracts, it is an antioxidant; the same chemical properties that give these substances their color makes them good antioxidants.

Berberine works to control obesity and diabetes by stimulating production of, or reducing the enzymatic destruction of, the incretin glucagon-like polypeptide 1 (GLP-1, see the section on hormones, page 112). This hormone increases insulin levels, decreases glucagon levels, and inhibits an enzyme that inhibits the insulin receptors in muscle and fat cells, thus increasing insulin sensitivity. Berberine also increases the burning of fatty acids and increases the number of insulin receptors in fat and muscle cells.

Because it is rapidly digested and cleared from the blood, it is given in frequent doses: 500 milligrams three times a day, preferably just after a meal. A three-month study of 36 diabetic patients comparing berberine to the antidiabetes drug metformin found that it had similar effects to that drug on fasting blood glucose levels, hemoglobin A1c, insulin sensitivity, and blood triglycerides (J. Yin, Xing, and Ye 2008). In the latter (blood fats and cholesterol) it worked better than metformin. Waist measurements and waist-to-hip ratio both declined.

Small studies for three months can't tell us much about the safety of a drug or herbal supplement. There may be unknown long-term effects on bones, pancreas, or liver cells, or harmful interactions with gut microbes. There is some evidence that berberine causes the liver to be less effective at clearing some drugs from the system. And, of course, herbal supplements are unregulated in the United States, so you can't really know if you are actually getting what you paid for. Drugs like metformin, on the other hand, have been subject to extensive long-term studies of their safety and effectiveness and are regulated.

Vinegar

Taking vinegar with food has long been known to reduce blood sugar after a meal. In the 1820s, Lord Byron ate a fad diet of potatoes drenched in vinegar to ward off obesity (he may have been anorexic). The active ingredient in vinegar is acetic acid, shown to the right.

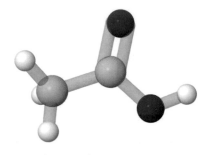

Fad diets aside, modern research into vinegar's antiglycemic properties bear it out (Johnston et al. 2010). Two tablespoons of vinegar (in a glass of water) taken during a carbohydrate meal (bagel or glucose) reduced blood sugar readings after the meal, compared to subjects taking no vinegar. The acidic form of vinegar (acetic acid) was required. Nonacidic sodium salts of acetic acid (sodium acetate) did not have any effect, thus showing that the effect was not simply due to the hormonal effects of acetate ions, which we saw earlier when looking at gut microbe activity (page 122).

Reminiscent of Lord Byron's diet of vinegar-soaked potatoes, one study on cold boiled potatoes with vinegar dressing found lower blood sugar levels and lower insulin levels after the meal (Leeman, Östman, and Björck 2005). Storing the potatoes in the refrigerator overnight allowed the amylose to crystallize into resistant starch. The vinegar slowed the digestion of the remaining starch.

Vinegar has been shown to slow the emptying of the stomach. This may be one of its mechanisms for reducing blood sugar levels, since it gives the carbohydrates time to be slowly digested and absorbed, without spiking blood sugar or insulin levels.

Vinegar also increases the satiety of a carbohydrate meal (Östman et al. 2005), in addition to reducing its effects on blood sugar and insulin. All these effects scaled in a linear fashion with how much vinegar was consumed with the meal: twice as much vinegar gave twice as much effect (up to the amounts tested, i.e., 30 milliliters [1 ounce] of white vinegar). Not only did vinegar increase satiety, it prolonged satiety, more than doubling it after two hours.

The effects of vinegar are similar to the effects of eating a low glycemic index meal compared to a high glycemic index meal. What vinegar seems to do is make those bad starches act more like a low glycemic index meal. This also means that if you are already eating low-carbohydrate meals, or low glycemic index meals, vinegar may not provide any benefit.

If you *must* eat that French bread at the restaurant, soak it in balsamic vinegar. But order the salad, with a vinaigrette dressing, and take your time enjoying the meal.

Vinegar not only reduces blood sugar and insulin during a meal, it also reduces insulin resistance (Johnston, Kim, and Buller 2004). That paper compared its effects to the prescription antidiabetes drug metformin. Vinegar interferes with an enzyme the body uses to digest disaccharides (sugars made up of two simple sugars) that result from the breakdown of starches. It also affects glucose metabolism in muscles, helping to reduce blood sugar levels and improve sensitivity to insulin. Similar results are found with other carbohydrates besides wheat, such as white rice, eaten either with vinegar or vinegar-pickled cucumber (Östman et al. 2005).

Other Dietary Considerations

Since overweight and obese people tend to underestimate how much they eat (Vliet-Ostaptchouk et al. 2008), just knowing this is the case and adjusting your estimate of calories or portion size can help you diet. If I always think 1,000 calories is 500, I can mentally double my estimate and think twice about how much I put on a plate.

EXERCISE

Exercise alone has been shown to convert fat-storing white fat into energy-burning beige fat. But exercising in temperatures around 60°F activates existing brown fat, raising metabolic rate and burning calories to produce heat.

Of course, exercising burns calories while you are doing it. Because of homeostasis mechanisms, this effect is tempered when the body reduces metabolic rate after the exercise to maintain body fat.

Exercise that builds muscle mass increases metabolic rate because all that new muscle needs to burn calories in order to live. Thus, building muscle raises metabolic rate permanently, while exercise that does not build muscle only increases metabolic rate while we are exercising.

Exercise relieves stress. Stress releases cortisol, which promotes fat storage. Exercise also reduces depression, which often contributes to obesity. Exercise suppresses appetite and helps you sleep better. All these benefits help you when you are trying to lose weight.

In one study specifically looking at chronic inflammation due to obesity, it was found that aerobic and resistance exercise reduces chronic inflammation in obese subjects (You et al. 2013). Besides its many impacts on general health, chronic inflammation also worsens obesity. Muscle cells produce anti-inflammatory molecules when they are exercised.

DRUGS

- Caffeine: Caffeine consumption raises adiponectin levels in the blood, reducing blood cholesterol levels and improving glucose metabolism. At the same time, it reduces leptin levels, and this reduces inflammation and blood triglycerides. It lowers the risk of type 2 diabetes and raises the levels of the good high-density lipoproteins (HDL) (Yamashita et al. 2012).
- Aspirin: Taking 600 milligrams of aspirin has been shown to be effective in reducing obesity-induced colon cancer rates (Movahedi et al. 2015). Aspirin is an anti-inflammatory, and inflammation exacerbates obesity.
- Metformin: The antidiabetes drug metformin has shown some mild weight-loss effects and helped some subjects keep from regaining weight. The effects were small, however.

- Ephedrine: Ephedrine is a stimulant whose use in weight loss has become illegal in the United States due to concern over its side effects. It has been shown to stimulate brown fat in mice, but in humans, chronic use actually reduces brown fat activity, despite being effective at reducing levels of white fat (Carey et al. 2015). In lean men, it activates brown fat, but in obese men, it does not (Carey et al. 2013). It slows the emptying of the stomach, which reduces hunger and slows down glucose uptake from carbohydrates. It is synergistic with caffeine and aspirin, meaning that the combination works better than any of the three drugs alone. As a drug, it is related to methamphetamine and can be used in the illicit manufacture of that illegal drug. Because of this, its availability is tightly regulated in the United States.
- Thiazolidinediones: These are drugs used in the treatment of diabetes. They activate brown fat, which burns calories to produce heat. They cause the body to store fatty acids in fat cells, causing a shortage of fatty acids in the blood, which causes an increase in the burning of glucose as the alternative fuel. They remove the fatty acids in the muscles that cause insulin resistance. They also reduce leptin, causing increased appetite, but they increase adiponectin, thus slowing the release of glucose from the liver. Unfortunately, most of these drugs have been removed from the market due to dangerous side effects.
- Mirabegron: This is a drug used to treat overactive bladder. It has been found to activate brown fat, in a similar way to how exposure to cold does.
- Orlistat: This is a prescription drug that prevents dietary fats from being absorbed during digestion. The effects are modest.

HORMONE THERAPY

Very few people have the rare genetic variants that disable the leptin or POMC systems. But for those who do, daily injections of recombinant leptin (or synthetic molecules that stimulate melanocortin-4 receptors) have effectively cured the syndrome, much as insulin injections have kept type 1 diabetics alive.

Thyroid hormones are also available for people with obesity due to hypothyroidism. But typically, changes in weight after treatment for hypothyroidism are modest. Thyroid deficiency is not often a contributing factor in obesity. Thyroid hormone levels already rise every time you eat, and hypothyroid condition is often caused by obesity (not the other way around) and goes away when the weight is lost (Longhi and Radetti 2013).

On the other hand, people being treated for hyperthyroid condition usually gain nontrivial amounts of weight. Thyroid hormone increases appetite, and people with hyperthyroidism are used to being able to eat large amounts of fattening foods without consequence, since they burn calories at a high rate.

Thyroid hormone therapy is not a safe or effective way to lose weight (Kaptein, Beale, and Chan 2009).

We discussed oxytocin in the chapter on hormones (page 120). Oxytocin nasal spray is available for purchase in online pharmacies and on Amazon.

REDUCING CHRONIC INFLAMMATION

Obesity causes chronic inflammation, and inflammation causes obesity. Reducing inflammation not only provides health benefits throughout the body, but it helps to reduce obesity.

There are many anti-inflammatory foods, and we have covered many of them already. We have also briefly mentioned aspirin as an anti-inflammatory aid, as well as the known anti-inflammatory effects of regular exercise.

Inflammation requires inflammatory molecules called eicosanoids. These are made from the fatty acid arachidonic acid (AA). The body makes arachidonic acid from linoleic acid in the diet. Up until about 50 years ago, the human diet had little linoleic acid in it. Traditional fats like lard, butter, and olive oil have less than 10 percent linoleic acid. But with the production of vegetable oils such as corn, soy, sunflower, and safflower oil, the linoleic acid content has risen to 50 percent and sometimes to 75 percent. These are inflammatory foods, and they are the cheapest source of fat, meaning that they are what goes into almost all processed food.

The other cheap source of calories is starch, particularly the high glycemic index starches found in wheat flour, rice, and potatoes. Insulin strongly activates key enzymes in the conversion of linoleic acid into inflammatory arachidonic acid. With sugar and starchy foods spiking insulin levels, and seed oils causing inflammation, our diet has evolved into one that promotes obesity. Only the genetically lucky few can eat these foods and stay lean.

Fighting inflammation with anti-inflammatory omega-3 oils, polyphenols, and plant pigments and reducing inflammatory oil consumption can go a long way to reducing or eliminating chronic inflammation. The omega-3 oil eicosapentaenoic acid (EPA) competes with arachidonic acid (AA) in cell membranes, preventing AA from being converted into inflammatory eicosanoids. It also inhibits one of the enzymes needed for the conversion of linoleic acid into arachidonic acid. EPA and another omega-3 oil, docosahexacnoic acid (DHA), activate genes that cause fat burning, thus reducing inflammatory blood fats (Sears and Ricordi 2010). These omega-3 oils are cheaper and safer than anti-inflammatory drugs such as aspirin and steroids. Many anti-inflammatory drugs work by preventing the conversion of AA into eicosanoids. Reducing AA and linoleic acid in the diet can have the same effects.

The two markers of inflammation are levels of C-reactive protein and the ratio of AA to EPA in the blood. If we lower the AA/EPA ratio by reducing AA and raising EPA, we will reduce inflammation and thus reduce C-reactive protein levels.

PSYCHOLOGICAL REMEDIES

Don't eat when you aren't hungry.

If you find yourself staring into an open refrigerator wondering what you want to eat, consider an apple or a carrot. If those do not appeal to you, you may not actually be hungry. You may be eating out of boredom, or looking for pleasure. Try going outside for a walk instead, or doing something else for fun, something that does not involve eating.

LIFESTYLE REMEDIES

Get more exercise. Find fun things to do that involve moving around, in situations away from food. Take a walk, ride a bike, play a sport, play with the dog, or do some gardening. There are lots of ways to have fun that involve moving parts of your body besides your jaw muscles.

Skip a meal. Or three or four. A day of fasting once a week can help you lose a pound a week. Do that for a year and you've lost 50 pounds.

Don't eat between meals. Give your body a few hours of low insulin levels to get back your insulin sensitivity. If you feel hungry, drink some water, coffee, tea, or thin broth. Change your attitude toward feeling hungry. Feeling hungry means you are making progress on losing weight. It makes the meal that follows all the more enjoyable.

Allow time for your body to realize you have eaten enough. Bolting a meal down in 15 minutes allows you to eat far more food than it takes to cure your hunger. It is what leads to a feeling of being overstuffed once your body has had time to register the amount of food that has just been dumped into it. Foods that take time to eat, such as most high-fiber foods (carrots, an apple, etc.), will slow you down at the table. Don't have second helpings until at least 20 minutes have gone by.

Keep high-fiber foods available and in plain sight, such as a bowl of fruit on the counter near the refrigerator. Get used to going to the fruit or vegetable drawer instead of to the other parts of the refrigerator.

Control portion size when you eat. Don't bring the whole pot to the table. Serve the food on plates in the kitchen and bring the plates to the dining table. If there is a big pile of food on the table, people will mindlessly eat it until it is gone. If they really need second helpings, a trip into the kitchen is not a hardship.

If you do want to have a big pile of food on the table, make it the high-fiber vegetables and salad. The meat, cheese, eggs, and starch are served on the plates and returned to the kitchen, and anyone who is still hungry will reach for the veggies and salad before they get up and run to the kitchen for another pork chop.

Don't bring sugar into the house. Don't allow the nondieters to bring it home either. They can find plenty of it elsewhere. When your sugar addiction has you running to the freezer for that pint of Ben & Jerry's, you will have to decide whether the drive to the store is really worth it.

Keep a food log. Some people like to take photos of everything they eat as a way of keeping track. Knowing how much you have eaten today, and what type of food it was, allows you to keep yourself honest.

Control inflammation by taking a baby aspirin every day (81 milli-grams) and by getting lots of colorful produce—that is how you know it contains antioxidants—and omega-3 fats, such as krill oil capsules.

Reduce stress by trying meditation or yoga. Exercise is also a great stress reliever. If there are stress-inducing parts of your life that you can do with-out, get rid of them.

SOCIAL REMEDIES

Do you overeat in social situations, like at a party or at the movies? Parties often provide lots of calorie-dense foods that stimulate the reward centers of your brain, even when you aren't hungry. The wide variety of foods at parties encourages overconsumption. We might get tired of eating one type of food, but when there are many different tastes available, we will continue to eat, long after we have stuffed ourselves.

To counter this, consider a decision not to eat anything at all at the party. Hold a glass of water in your hand and take a sip whenever you are tempted to grab a handful of chips or candy. If the party isn't fun without overeating, you are going to the wrong parties. Explain to people that you are planning not to eat anything, so they will help you with your willpower. If they don't help, or try to talk you out of it, they are not the kind of people you want to be around when you are trying to lose weight.

At the movies, simply avoid the concession stand. You will save a lot of money. If your only reason for going to the movie was to gorge on popcorn and candy, perhaps a movie is not really your favorite form of entertainment.

Many people think that in order to be a good host, they must feed their guests, and feed them party foods instead of healthy fare. But your guests will still come over if you explain to them beforehand that food will not be served. They will eat before they come or make plans to eat afterward. If you want to hedge your bets, have some nuts and bowls of fruit set around. Those who are genuinely hungry will have something healthy to eat. If an apple doesn't make them happy, they weren't actually hungry.

At social gatherings, people tend to eat mindlessly, simply because there is food within easy reach everywhere. An engaging conversation or a good movie makes people forget that they aren't really hungry. If the food isn't around, they won't eat, and they won't miss it.

If the gathering is actually about food, bring the food to a table from the kitchen, where it was out of sight. When the meal is finished, remove the food, or move the gathering away from the food, so people don't continuously graze mindlessly. Instead of having snacks placed everywhere to eat before the meal, plan to have the meal soon after people arrive, and move away from the table, or move the food, when the meal is over. Plan to have enough food to appease hunger, and no more, so there isn't a lot of food lying around when people are no longer hungry.

VICIOUS CIRCLES

Throughout this book, I have mentioned harmful positive feedback effects, where being in a particular state, such as obesity, has effects on other systems in the body that then feed back and amplify the original problem. Here is a list of those "vicious circles":

- Obesity causes poor sleep and poor sleep causes obesity.
- Visceral fat reduces growth hormone, leading to more visceral fat.
- Since obesity causes insulin resistance, and insulin resistance causes leptin resistance, once we become obese, we are always hungry and become more obese.
- Stress may cause obesity, which causes more stress.

- High-sugar, high-fat diet increases Firmicutes bacteria in the gut, leading to increased energy absorption, exacerbating the weight gain from the poor diet.
- Obese mothers program their fetuses to become obese, producing a new generation of obese mothers.
- Fat cells cause autoimmune attacks on the thyroid gland, reducing thyroid hormones, further contributing to obesity.
- A high-fat diet causes the intestinal K cells that produce the hormone GIP to grow and proliferate. GIP allows fat cells to take up fatty acids and get fatter.
- Depression can lead to overeating and weight gain, and obesity can lead to depression.
- Western diet increases fat-inducing bacteria that produce toxic DCA, which kills other bacteria, allowing the fat-inducing ones to take over.
- Inflammation increases AGRP, which increases inflammation.

BIBLIOGRAPHY

Abraham, S. B., et al. 2013. "Cortisol, Obesity and the Metabolic Syndrome: A Cross-Sectional Study of Obese Subjects and Review of the Literature." *Obesity* 21 (1): E105–E117. http://dx.doi.org/10.1002/oby.20083.

Ameri, Pietro, et al. 2013. "Vitamin D Increases Circulating IGF1 in Adults: Potential Implication for the Treatment of GH Deficiency." *European Journal of Endocrinology* 168 (6): 767–772. http://dx.doi.org/10.1530/EJE-13-0510.

Anson, R. Michael, et al. 2002. "Intermittent Fasting Dissociates Beneficial Effects of Dietary Restriction on Glucose Metabolism and Neuronal Resistance to Injury from Calorie Intake." *Proceedings of the National Academy of Sciences of the United States of America* 100 (10): 6216–6220. http://dx.doi.org/10.1073/pnas.1035720100.

Arafat, A. M., et al. 2013. "The Impact of Insulin-Independent, Glucagon-Induced Suppression of Total Ghrelin on Satiety in Obesity and Type 1 Diabetes Mellitus." *Journal of Clinical Endocrinology and Metabolism* 98 (10). http://dx.doi.org/10.1210/jc.2013-1635.

Arumagam, Manimozhiyan, et al. 2011. "Enterotypes of the Human Gut Microbiome." *Nature* 473:174–180. http://dx.doi.org/10.1038/nature09944.

Astrup, Arne. 2008. "The Role of Calcium in Energy Balance and Obesity: The Search for Mechanisms." *American Journal of Clinical Nutrition* 88 (4): 873–874. http://ajcn.nutrition.org/content/88/4/873.full.pdf.

Astrup, Arne, A. Raben, and Nina Geiker. 2015. "The Role of Higher Protein Diets in Weight Control and Obesity-Related Comorbidities." *International Journal of Obesity* 39:721–726. http://dx.doi.org/10.1038/ijo.2014.216.

Avena, Nicole M., and Bartley G. Hoebel. 2003. "A Diet Promoting Sugar Dependency Causes Behavioral Cross-Sensitization to a Low Dose of Amphetamine." *Neuroscience* 122 (1): 17–20. http://dx.doi.org/10.1016/S0306-4522(03)00502-5.

Avena, Nicole M., Pedro Rada, and Bartley G. Hoebel. 2008. "Evidence for Sugar Addiction: Behavioral and Neurochemical Effects of Intermittent, Excessive Sugar Intake." *Neuroscience & Biobehavioral Reviews* 32 (1): 20–39. http://dx.doi.org/10.1016/j.neubiorev.2007.04.019.

Bäckhed, Fredrik, et al. 2004. "The Gut Microbiota as an Environmental Factor That Regulates Fat Storage." *Proceedings of the National Academy of Sciences of the United States of America* 101 (44): 15718–15723. http://dx.doi.org/10.1073/pnas.0407076101.

Baker, Monya. 2017. "Deceptive Curcumin Offers Cautionary Tale for Chemists." *Nature* 541:144–145. https://dx.doi.org/10.1038/541144a.

Barsh, Gregory S., I. Sadaf Farooqi, and Stephen O'Rahilly. 2000. "Genetics of Body-Weight Regulation." *Nature* 404:644–651. https://dx.doi.org/10.1038/35007519.

Beck, B. 2006. "Neuropeptide Y in Normal Eating and in Genetic and Dietary-Induced Obesity." *Philosophical Transactions of the Royal Society B* 361 (1471): 1159–1185. https://dx.doi.org/10.1098%2Frstb.2006.1855.

Behall, K. M., and J. Hallfrisch. 2002. "Plasma Glucose and Insulin Reduction After Consumption of Breads Varying in Amylose Content." *European Journal of Clinical Nutrition* 56 (9): 913–920. http://dx.doi.org/10.1038/sj.ejcn.1601411.

Bernstein, Adam M., Michael F. Roizen, and Luis Martinez. 2014. "Purified Palmitoleic Acid for the Reduction of High-Sensitivity C-Reactive Protein and Serum Lipids: A Double-Blinded, Randomized, Placebo Controlled Study." *Journal of Clinical Lipidology* 8 (6): 612–617. http://dx.doi.org/10.1016/j.jacl.2014.08.001.

Bhattacharya, Arunabh, et al. 2005. "The Combination of Dietary Conjugated Linoleic Acid and Treadmill Exercise Lowers Gain in Body Fat Mass and Enhances Lean Body Mass in High Fat-Fed Male Balb/C Mice." *Journal of Nutrition* 135 (5): 1124–1130. http://jn.nutrition.org/content/135/5/1124.long.

Bhattacharya, Arunabh, et al. 2006. "Biological Effects of Conjugated Linoleic Acids in Health and Disease." *Journal of Nutritional Biochemistry* 17 (12): 789–810. https://dx.doi.org/10.1016/j.jnutbio.2006.02.009.

Biondi, Bernadette. 2010. "Thyroid and Obesity: An Intriguing Relationship." *Journal of Clinical Endocrinology & Metabolism* 95 (8): 3614–3617. http://dx.doi.org/10.1210/jc.2010-1245.

Blázquez, Enrique, et al. 2014. "Insulin in the Brain: Its Pathophysiological Implications for States Related with Central Insulin Resistance, Type 2 Diabetes and Alzheimer's Disease." *Frontiers in Endocrinology* 5:161. http://dx.doi.org/10.3389/fendo.2014.00161.

Bose, Mousumi, Blanca Oliván, and Blandine Laferrère. 2009. "Stress and Obesity: The Role of the Hypothalamic–Pituitary–Adrenal Axis in Metabolic Disease." *Current Opinion in Endocrinology, Diabetes and Obesity* 16 (5): 340–346. https://dx.doi.org/10.1097%2FMED.0b013e32832fa137.

Bostrom, Pontus, et al. 2012. "A PGC1-a-Dependent Myokine That Drives Brown-Fat-Like Development of White Fat and Thermogenesis." *Nature* 481 (7382): 463–469. http://dx.doi.org/10.1038/nature10777.

Bougnères, P., et al. 1997. "In Vivo Resistance of Lipolysis to Epinephrine. A New Feature of Childhood Onset Obesity." *Journal of Clinical Investigation* 99 (11): 2568–2573. http://dx.doi.org/10.1172/JCI119444.

Bourzac, Katherine. 2014. "The Bacterial Tightrope." *Nature* 516:S14–S16. https://dx.doi.org/10.1038/516S14a.

Bowers, Cyril Y. 2012. "Ghrelin: A History of Its Discovery." In *Ghrelin in Health and Disease*, edited by Roy G. Smith and Michael O. Thorner, 1–35. New York: Humana Press. http://dx.doi.org/10.1016/B978-0-12-381272-8.00001-5.

Breier, B. H., et al. 2001. "Fetal Programming of Appetite and Obesity." *Molecular and Cellular Endocrinology* 185 (1–2): 73–79. http://dx.doi.org/10.1016/S0303-7207(01)00634-7.

Byrnes, S. E., J. C. Miller, and G. S. Denyer. 1995. "Amylopectin Starch Promotes the Development of Insulin Resistance in Rats." *Journal of Nutrition* 125 (6): 1430–1437. www.ncbi.nlm.nih.gov/pubmed/7782895.

Campbell, Kristin L., et al. 2012. "Reduced-Calorie Dietary Weight Loss, Exercise, and Sex Hormones in Postmenopausal Women: Randomized Controlled Trial." *Journal of Clinical Oncology* 30 (19): 2314–2326. http://dx.doi.org/10.1200/JCO.2011 .37.9792.

Cano, Paola Gauffin, et al. 2012. "*Bacteroides uniformis* CECT 7771 Ameliorates Metabolic and Immunological Dysfunction in Mice with High-Fat-Diet Induced Obesity." *PLoS One* 7 (7): e41079. http://dx.doi.org/10.1371/journal.pone.0041079.

Cappelleri, J. C., et al. 2009. "Evaluating the Power of Food Scale in Obese Subjects and a General Sample of Individuals: Development and Measurement Properties." *International Journal of Obesity* 33:913–922. http://dx.doi.org/10.1038/ijo.2009.107.

Cappon, J. P., et al. 2009. "Acute Effects of High Fat and High Glucose Meals on the Growth Hormone Response to Exercise." *Journal of Clinical Endocrinology and Metabolism* 76 (6): 1418–1422. http://dx.doi.org/10.1210/jcem.76.6.8501145.

Carey, Andrew L., et al. 2013. "Ephedrine Activates Brown Adipose Tissue in Lean but Not Obese Humans." *Diabetologia* 56 (1): 147–155. https://dx.doi.org/10.1007/s00125 -012-2748-1.

Carey, Andrew L., et al. 2015. "Chronic Ephedrine Administration Decreases Brown Adipose Tissue Activity in a Randomised Controlled Human Trial: Implications for Obesity." *Diabetologia* 58 (5): 1045–1054. http://dx.doi.org/10.1007/s00125 -015-3543-6.

Carlsson, Björn, et al. 1997. "Obese (ob) Gene Defects Are Rare in Human Obesity." *Obesity: A Research Journal* 5 (1): 30–35. http://dx.doi.org/10.1002/j.1550-8528 .1997.tb00280.x.

Carulli, L., et al. 1999. "Regulation of *ob* Gene Expression: Evidence for Epinephrine-Induced Suppression in Human Obesity." *Journal of Clinical Endocrinology and Metabolism* 84 (9): 3309–3312. http://dx.doi.org/10.1210/jcem.84.9.6007.

Chaix, Amandine, et al. 2014. "Time-Restricted Feeding Is a Preventative and Therapeutic Intervention Against Diverse Nutritional Challenges." *Cell Metabolism* 20 (6): 991–1005. http://dx.doi.org/10.1016/j.cmet.2014.11.001.

Chandalia, Manisha, et al. 2000. "Beneficial Effects of High Dietary Fiber Intake in Patients with Type 2 Diabetes Mellitus." *New England Journal of Medicine* 342 (19): 1392–1398. http://dx.doi.org/10.1056/NEJM200005113421903.

Chandarana, Keval, and Rachel L. Batterham. 2012. "Shedding Pounds After Going Under the Knife: Metabolic Insights from Cutting the Gut." *Nature Medicine* 18:668–669. http://dx.doi.org/doi:10.1038/nm.2748.

Charradi, Kamel, et al. 2011. "Grape Seed Extract Alleviates High-Fat Diet-Induced Obesity and Heart Dysfunction by Preventing Cardiac Siderosis." *Cardiovascular Toxicology* 11 (1): 28–37. https://dx.doi.org/10.1007/s12012-010-9101-z.

Chatelier, Emmanuelle Le, et al. 2013. "Richness of Human Gut Microbiome Correlates with Metabolic Markers." *Nature* 500:541–546. http://dx.doi.org/10.1038 /nature12506.

Chen, Xiaoping, and Wenying Yang. 2015. "Branched-Chain Amino Acids and the Association with Type 2 Diabetes." *Journal of Diabetes Investigation* 6 (4): 369–370. https://dx.doi.org/10.1111%2Fjdi.12345.

Chondronikola, Maria, et al. 2014. "Brown Adipose Tissue Improves Whole-Body Glucose Homeostasis and Insulin Sensitivity in Humans." *Diabetes* 63 (12): 4089–4099. http://dx.doi.org/10.2337/db14-0746.

Christensen, Mikkel, et al. 2011. "Glucose-Dependent Insulinotropic Polypeptide." *Diabetes* 60 (12): 3103–3109. http://dx.doi.org/10.2337/db11-0979.

Clasey, Jody L., et al. 2001. "Abdominal Visceral Fat and Fasting Insulin Are Important Predictors of 24-Hour GH Release Independent of Age, Gender, and Other Physiological Factors." *Journal of Clinical Endocrinology and Metabolism* 86 (8): 3845–3852. http://dx.doi.org/10.1210/jcem.86.8.7731.

Claussnitzer, Melina, et al. 2015. "FTO Obesity Variant Circuitry and Adipocyte Browning in Humans." *New England Journal of Medicine* 373 (10): 895–907. http://dx.doi.org/10.1056/NEJMoa1502214.

Coakley, M., et al. 2002. "Conjugated Linoleic Acid Biosynthesis by Human-Derived Bifidobacterium Species." *Journal of Applied Microbiology* 94 (1): 138–145. http://dx.doi.org/10.1046/j.1365-2672.2003.01814.x.

Cotillard, Aurélie, et al. 2013. "Dietary Intervention Impact on Gut Microbial Gene Richness." *Nature* 500:585–588. https://dx.doi.org/10.1038/nature12480.

Cypess, Aaron M., et al. 2012. "Cold but Not Sympathomimetics Activates Human Brown Adipose Tissue In Vivo." *Proceedings of the National Academy of Sciences* 109 (25): 10001–10005. http://dx.doi.org/10.1073/pnas.1207911109.

Cypess, Aaron M., et al. 2015. "Activation of Human Brown Adipose Tissue by a β3-Adrenergic Receptor Agonist." *Cell Metabolism* 21 (1): 33–38. http://dx.doi.org/10.1016/j.cmet.2014.12.009.

Dallman, Mary F., et al. 2003. "Chronic Stress and Obesity: A New View of 'Comfort Food.'" *Proceedings of the National Academy of Sciences* 100 (20): 11696–11701. www.researchgate.net/profile/Susanne_La_Fleur/publication/10565407_Chronic_stress_and_obesity_a_new_view_of_comfort_food/links/004635203c9af65405000000.pdf.

Dandona, Paresh, Ahmad Aljada, and Arindam Bandyopadhyay. 2004. "Inflammation: The Link Between Insulin Resistance, Obesity and Diabetes." *TRENDS in Immunology* 25 (1): 4–7. http://dx.doi.org/10.1016/j.it.2003.10.013.

Dao, Maria Carlota, et al. 2014. "*Akkermansia muciniphila* and Improved Metabolic Health During a Dietary Intervention in Obesity: Relationship with Gut Microbiome Richness and Ecology." *Gut* 65:426–436. http://dx.doi.org/10.1136/gutjnl-2014-308778.

Davis, Caroline. 2013. "From Passive Overeating to 'Food Addiction': A Spectrum of Compulsion and Severity." *ISRN Obesity* 2013. http://dx.doi.org/10.1155/2013/435027.

Davis, Caroline A., et al. 2009. "Dopamine for 'Wanting' and Opioids for 'Liking': A Comparison of Obese Adults with and Without Binge Eating." *Obesity* 17 (6): 1220–1225. http://dx.doi.org/10.1038/oby.2009.52.

Deblon, Nicolas, et al. 2011. "Mechanisms of the Anti-obesity Effects of Oxytocin in Diet-Induced Obese Rats." *PLoS One* 6 (9): e25565. http://dx.doi.org/10.1371/journal.pone.0025565.

Dirlewanger, M., et al. 2000. "Effects of Short-Term Carbohydrate or Fat Overfeeding on Energy Expenditure and Plasma Leptin Concentrations in Healthy Female Subjects." *International Journal of Obesity and Related Metabolic Disorders* 24 (11): 1413–1418. www.ncbi.nlm.nih.gov/pubmed/11126336.

Dulloo, Abdul G., et al. 1999. "Efficacy of a Green Tea Extract Rich in Catechin Polyphenols and Caffeine in Increasing 24-h Energy Expenditure and Fat Oxidation in Humans." *American Society for Clinical Nutrition* 70 (6): 1040–1045. http://ajcn.nutrition.org/content/70/6/1040.full.pdf+html.

Duncan, Sylvia H., et al. 2006. "Reduced Dietary Intake of Carbohydrates by Obese Subjects Results in Decreased Concentrations of Butyrate and Butyrate-Producing Bacteria in Feces." *Applied and Environmental Microbiology* 73 (4): 1073–1078. https://dx.doi.org/10.1128/AEM.02340-06.

Ebbeling, Cara B., et al. 2007. "Effects of a Low–Glycemic Load vs. Low-Fat Diet in Obese Young Adults." *Journal of the American Medical Association* 297 (19): 2092–2102. http://jama.jamanetwork.com/article.aspx?articleid=207088.

Ebbeling, Cara B., et al. 2012. "Effects of Dietary Composition on Energy Expenditure During Weight-Loss Maintenance." *Journal of the American Medical Association* 307 (24): 2627–2634. http://dx.doi.org/10.1001/jama.2012.6607.

Egecioglu, Emil, et al. 2010. "Ghrelin Increases Intake of Rewarding Food in Rodents." *Addiction Biology* 15 (3): 304–311. https://dx.doi.org/10.1111%2Fj.1369-1600.2010 .00216.x.

Eiseman, B., et al. 1958. "Fecal Enema as an Adjunct in the Treatment of Pseudomembranous Enterocolitis." *Surgery* 44 (5): 854–859. www.ncbi.nlm.nih.gov /pubmed/13592638.

Epel, Elissa, et al. 2001. "Stress May Add Bite to Appetite in Women: A Laboratory Study of Stress-Induced Cortisol and Eating Behavior." *Psychoneuroendocrinology* 26 (1): 37–49. http://dx.doi.org/10.1016/S0306-4530(00)00035-4.

Everard, Amandine, et al. 2014. "Saccharomyces boulardii Administration Changes Gut Microbiota and Reduces Hepatic Steatosis, Low-Grade Inflammation, and Fat Mass in Obese and Type 2 Diabetic db/db mice." *mBio* 5 (3): e01011 e01014. http://dx.doi .org/10.1128/mBio.01011-14.

Fehm, Horst L., et al. 2001. "The Melanocortin Melanocyte-Stimulating Hormone/ Adrenocorticotropin4–10 Decreases Body Fat in Humans." *Journal of Clinical Endocrinology & Metabolism* 86:1144–1148. http://dx.doi.org/10.1210/jcem.86.3 .7298.

Fleurbaix Laventie Ville Sante (FLVS) Study Group. 2004. "The Three-Factor Eating Questionnaire-R18 Is Able to Distinguish Among Different Eating Patterns in a General Population." *Journal of Nutrition* 134 (9): 2372–2380. http://jn.nutrition.org /content/134/9/2372.full.

Fraser, Robert, et al. 1999. "Cortisol Effects on Body Mass, Blood Pressure, and Cholesterol in the General Population." *Hypertension* 33 (6): 1364–1368. http:// dx.doi.org/10.1161/01.HYP.33.6.1364.

Friedman, Jeffrey M. 2010. "A Tale of Two Hormones." *Nature Medicine* 16 (10): 1100– 1106. https://dx.doi.org/10.1038/nm1010-1100.

Gamber, Kevin M., Heather Macarthur, and Thomas C. Westfall. 2005. "Cannabinoids Augment the Release of Neuropeptide Y in the Rat Hypothalamus." *Neuropharmacology* 49 (5): 646–652. http://dx.doi.org/10.1016/j.neuropharm.2005 .04.017.

Gonzalez-Rey, Elena, Alejo Chorny, and Mario Delgado. 2006. "Therapeutic Action of Ghrelin in a Mouse Model of Colitis." *Gastroenterology* 130 (6): 1707–1720. https:// dx.doi.org/10.1053%2Fj.gastro.2006.01.041.

Goodrich, Julia K., et al. 2014. "Human Genetics Shape the Gut Microbiome." *Cell* 159 (4): 789–799. http://dx.doi.org/10.1016/j.cell.2014.09.053.

Granfeldt, Y., A. Drews, and I. Björck. 1995. "Arepas Made from High Amylose Corn Flour Produce Favorably Low Glucose and Insulin Responses in Healthy Humans." *Journal of Nutrition* 125 (3): 459–465. www.ncbi.nlm.nih.gov/pubmed/7876921.

Grantham, James P., and Maciej Henneberg. 2014. "The Estrogen Hypothesis of Obesity." *PLoS One* 9 (6): e99776. http://dx.doi.org/10.1371/journal.pone.0099776.

Green, Courtney R., et al. 2016. "Branched-Chain Amino Acid Catabolism Fuels Adipocyte Differentiation and Lipogenesis." *Nature Chemical Biology* 12:15–21. http://dx.doi.org/10.1038/nchembio.1961.

Greenberg, Andrew S., and Martin S. Obin. 2006. "Obesity and the Role of Adipose Tissue in Inflammation and Metabolism." *American Journal of Clinical Nutrition* 83 (2): 461S–465S. http://ajcn.nutrition.org/content/83/2/461S.full.pdf+html.

Günther, Anke L. B., et al. 2007. "Early Protein Intake and Later Obesity Risk: Which Protein Sources at Which Time Points Throughout Infancy and Childhood Are Important for Body Mass Index and Body Fat Percentage at 7 y of Age?" *American Journal of Clinical Nutrition* 86 (6): 1765–1772. http://ajcn.nutrition.org/content/86/6/1765.full.pdf.

Gutierrez-Aguilar, Ruth, and Stephen C. Woods. 2011. "Nutrition and L and K-enteroendocrine Cells." *Current Opinion in Endocrinology, Diabetes and Obesity* 18 (1): 35–41. https://dx.doi.org/10.1097%2FMED.0b013e32834190b5.

Halberg, Nils, et al. 2005. "Effect of Intermittent Fasting and Refeeding on Insulin Action in Healthy Men." *Journal of Applied Phisiology* 99:2128–2136. http://dx.doi.org/10.1152/japplphysiol.00683.2005.

Hankey, Catherine, Dominika Klukowska, and Michael Lean. 2015. "A Systematic Review of the Literature on Intermittent Fasting for Weight Management." *FASEB Journal* 29 (suppl. 1): 117.4. http://dx.doi.org/10.1096/fj.1530-6860.

Harms, Matthew, and Patrick Seale. 2012. "Brown and Beige Fat: Development, Function and Therapeutic Potential." *Nature Medicine* 19:1252–1263. http://dx.doi.org/10.1038/nm.3361.

Harris, Ruth B. S. 2014. "Direct and Indirect Effects of Leptin on Adipocyte Metabolism." *Biochimica et Biophysica Acta (BBA)—Molecular Basis of Disease* 1842 (3): 414–423. http://dx.doi.org/10.1016/j.bbadis.2013.05.009.

Harrold, Joanne A., and Gareth Williams. 2003. "The Cannabinoid System: A Role in Both the Homeostatic and Hedonic Control of Eating?" *British Journal of Nutrition* 90 (4): 729–734. https://doi.org/10.1079/BJN2003942.

Hartman, M. L., et al. 1992. "Augmented Growth Hormone (GH) Secretory Burst Frequency and Amplitude Mediate Enhanced GH Secretion During a Two-Day Fast in Normal Men." *Journal of Clinical Endocrinology and Metabolism* 74 (4): 757–765. http://dx.doi.org/10.1210/jcem.74.4.1548337.

Harvie, M. N., et al. 2010. "The Effects of Intermittent or Continuous Energy Restriction on Weight Loss and Metabolic Disease Risk Markers: A Randomized Trial in Young Overweight Women." *International Journal of Obesity* 35:714–727. http://dx.doi.org/10.1038/ijo.2010.171.

Hervey, G. R. 1959. "The Effects of Lesions in the Hypothalamus in Parabiotic Rats." *Journal of Physiology* 145:336–352. www.ncbi.nlm.nih.gov/pmc/articles/PMC1356830/pdf/jphysiol01309-0116.pdf.

Ho, Klan Y., et al. 1988. "Fasting Enhances Growth Hormone Secretion and Amplifies the Complex Rhythms of Growth Hormone Secretion in Man." *Journal of Clinical Investigation*, 81 (4): 968–975. https://dx.doi.org/10.1172%2FJCI113450.

Howarth, Nancy C., Edward Saltzman, and Susan B. Roberts. 2001. "Dietary Fiber and Weight Regulation." *Nutrition Reviews* 59 (5): 129–139. http://dx.doi.org/10.1111/j.1753-4887.2001.tb07001.x.

Hsu, Chin-Lin, et al. 2008. "Phenolic Compounds Rutin and o-Coumaric Acid Ameliorate Obesity Induced by High-Fat Diet in Rats." *Journal of Agricultural and Food Chemistry* 57 (2): 425–431. http://dx.doi.org/10.1021/jf802715t.

Hu, Tian, et al. 2012. "Effects of Low-Carbohydrate Diets Versus Low-Fat Diets on Metabolic Risk Factors: A Meta-analysis of Randomized Controlled Clinical Trials." *American Journal of Epidemiology* 176 (suppl. 7): S44–S54. http://aje.oxfordjournals.org/content/176/suppl_7/S44.full.

Huynh, Frank K., et al. 2012. "A Role for Hepatic Leptin Signaling in Lipid Metabolism via Altered Very Low Density Lipoprotein Composition and Liver Lipase Activity in Mice." *Hepatology* 57 (2): 543–554. https://dx.doi.org/10.1002/hep.26043.

Ingalls, Ann M., Margaret M. Dickie, and G. D. Snell. 1950. "Obese, a New Mutation in the House Mouse." *Journal of Heredity* 41 (12): 317–318. www.oxfordjournals.org /our_journals/jhered/freepdf/41-317.pdf.

Israel, Y., et al. 2005. "NPY-Induced Feeding: Pharmacological Characterization Using Selective Opioid Antagonists and Antisense Probes in Rats." *Peptides* 26 (7): 1167–1175. http://dx.doi.org/10.1016/j.peptides.2005.01.017.

Jang, Cholsoon, et al. 2016. "A Branched Chain Amino Acid Metabolite Drives Vascular Transport of Fat and Causes Insulin Resistance." *Nature Medicine* 22 (4): 421–426. https://dx.doi.org/10.1038/nm.4057.

Jankowski, Marck, Tom L. Broderick, and Jolanta Gutkowska. 2016. "Oxytocin and Cardioprotection in Diabetes and Obesity." *BMC Endocrine Disorders* 16:34. http:// dx.doi.org/10.1186/s12902-016-0110-1.

Janssens, Pilou L. H. R., et al. 2013. "Acute Effects of Capsaicin on Energy Expenditure and Fat Oxidation in Negative Energy Balance." *PLoS One* 8 (7): e67786. http:// dx.doi.org/10.1371/journal.pone.0067786.

Jéquier, Eric. 1999. "Alcohol Intake and Body Weight: A Paradox." *American Journal of Clinical Nutrition* 69 (2): 173–174. http://ajcn.nutrition.org/content/69/2/173.full.

Johnson, Richard J., et al. 2007. "Potential Role of Sugar (Fructose) in the Epidemic of Hypertension, Obesity and the Metabolic Syndrome, Diabetes, Kidney Disease, and Cardiovascular Disease." *American Journal of Clinical Nutrition* 86 (4): 899–906. http://ajcn.nutrition.org/content/86/4/899.long.

Johnston, Carol S., Cindy M. Kim, and Amanda J. Buller. 2004. "Vinegar Improves Insulin Sensitivity to a High-Carbohydrate Meal in Subjects with Insulin Resistance or Type 2 Diabetes." *Diabetes Care* 27 (1): 281–282. http://dx.doi.org/10.2337/diacare .27.1.281.

Johnston, Carol S., et al. 2010. "Examination of the Antiglycemic Properties of Vinegar in Healthy Adults." *Annals of Nutrition and Metabolism* 56 (1): 74–79. https://dx.doi .org/10.1159/000272133.

Kadooka, Y., et al. 2010. "Regulation of Abdominal Adiposity by Probiotics (Lactobacillus gasseri SBT2055) in Adults with Obese Tendencies in a Randomized Controlled Trial." *European Journal of Clinical Nutrition* 64:636–643. http://dx.doi.org/10.1038 /ejcn.2010.19.

Kahn, Barbara B., and Jeffrey S. Flier. 2000. "Obesity and Insulin Resistance." *Journal of Clinical Investigation* 106 (4): 473–481. http://dx.doi.org/10.1172/JCI10842.

Kalra, Staya P., and Pushpa S. Kalra. 2003. "A Physiological Orexigen Modulated by the Feedback Action of Ghrelin and Leptin." *Endocrine* 22:49-55. http://dx.doi .org/10.1385/ENDO:22:1:49.

Kang, Ji-Hye, et al. 2010. "Dietary Capsaicin Reduces Obesity-Induced Insulin Resistance and Hepatic Steatosis in Obese Mice Fed a High-Fat Diet." *Obesity* 18 (4): 780–787. http://dx.doi.org/10.1038/oby.2009.301.

Kang, Young-Rye, et al. 2012. "Anti-Obesity and Anti-Diabetic Effects of Yerba Mate (Ilex paraguariensis) in C57BL/6J Mice Fed a High-Fat Diet." *Laboratory Animal Research* 28 (1): 23–29. https://dx.doi.org/10.5625%2Flar.2012.28.1.23.

Kaptein, Elaine M., Elizabeth Beale, and Linda S. Chan. 2009. "Thyroid Hormone Therapy for Obesity and Nonthyroidal Illnesses: A Systematic Review." *Journal of Clinical Endocrinology and Metabolism* 94 (10): 3663–3675. https://dx.doi .org/10.1210/jc.2009-0899.

Keller, Ulrich. 2011. "Dietary Proteins in Obesity and in Diabetes." *International Journal for Vitamin and Nutrition Research* 81 (2–3): 125–133. https://dx.doi.org/10.1024/0300-9831/a000059.

Kendall, A., et al. 1991. "Weight Loss on a Low-Fat Diet: Consequence of the Imprecision of the Control of Food Intake in Humans." *American Society for Clinical Nutrition* 53 (5): 1124–1129. http://ajcn.nutrition.org/content/53/5/1124.short.

Kim, K. R., et al. 1999. "Low-Dose Growth Hormone Treatment with Diet Restriction Accelerates Body Fat Loss, Exerts Anabolic Effect and Improves Growth Hormone Secretory Dysfunction in Obese Adults." *Hormone Research in Paediatrics* 51 (2): 78–84. http://dx.doi.org/10.1159/000023319.

King, N. A., et al. 2012. "Exercise, Appetite and Weight Management: Understanding the Compensatory Responses in Eating Behaviour and How They Contribute to Variability in Exercise-Induced Weight Loss." *British Journal of Sports Medicine* 46:315–322. http://dx.doi.org/10.1136/bjsm.2010.082495.

Kojima, M., and K. Kangawa. 2013. "Ghrelin Discovery: A Decade After." In *The Ghrelin System*, edited by A. Benso et al., 1–4. Turin, Italy: Karger. http://dx.doi.org/10.1159%2F000346036.

Koletzko, Berthold, et al. 2016. "High Protein Intake in Young Children and Increased Weight Gain and Obesity Risk." *American Journal of Clinical Nutrition* 103 (2): 303–304. http://dx.doi.org/10.3945/ajcn.115.128009.

Koska, Juraj, et al. 2004. "Pancreatic Polypeptide Is Involved in the Regulation of Body Weight in Pima Indian Male Subjects." *Diabetes* 53 (12): 3091–3096. http://dx.doi.org/10.2337/diabetes.53.12.3091.

Kovatcheva-Datchary, Petia, et al. 2015. "Dietary Fiber–Induced Improvement in Glucose Metabolism Is Associated with Increased Abundance of Prevotella." *Cell Metabolism* 22 (6): 971–982. http://dx.doi.org/10.1016/j.cmet.2015.10.001.

Larsson, Susanna C., Leif Bergkvist, and Alicja Wolk. 2006. "Processed Meat Consumption, Dietary Nitrosamines and Stomach Cancer Risk in a Cohort of Swedish Women." *Journal of Cachexia, Sarcopenia and Muscle* 119 (4): 915–919. http://dx.doi.org/10.1002/ijc.21925.

Lartigue, Guillaume de, et al. 2007. "Cocaine- and Amphetamine-Regulated Transcript: Stimulation of Expression in Rat Vagal Afferent Neurons by Cholecystokinin and Suppression by Ghrelin." *Journal of Neuroscience* 27 (11): 2876–2882. http://dx.doi.org/10.1523/JNEUROSCI.5508-06.2007.

Lee, Gong-Rak, et al. 2013. "Topical Application of Capsaicin Reduces Visceral Adipose Fat by Affecting Adipokine Levels in High-Fat Diet-Induced Obese Mice." *Obesity* 21 (1): 115–122. https://dx.doi.org/10.1002/oby.20246.

Lee, Mak-Soon, et al. 2011. "Reduction of Body Weight by Dietary Garlic Is Associated with an Increase in Uncoupling Protein mRNA Expression and Activation of AMP-Activated Protein Kinase in Diet-Induced Obese Mice." *Journal of Nutrition* 141 (11): 1947–1953. http://dx.doi.org/10.3945/jn.111.146050.

Lee, Zoe S. K., et al. 2001. "Urinary Epinephrine and Norepinephrine Interrelations with Obesity, Insulin, and the Metabolic Syndrome in Hong Kong Chinese." *Metabolism Clinical and Experimental* 50 (2): 135–143. https://dx.doi.org/10.1053/meta.2001.19502.

Leeman, M., E. Östman, and I. Björck. 2005. "Vinegar Dressing and Cold Storage of Potatoes Lowers Postprandial Glycaemic and Insulinaemic Responses in Healthy Subjects." *European Journal of Clinical Nutrition* 59 (11): 1266–1271. https://dx.doi.org/10.1038/sj.ejcn.1602238.

Leung, Felix W. 2008. "Capsaicin-Sensitive Intestinal Mucosal Afferent Mechanism and Body Fat Distribution." *Life Sciences* 83 (1–2): 1–5. http://dx.doi.org/10.1016/j .lfs.2008.04.018.

Leung, Felix W. 2014. "Capsaicin as an Anti-Obesity Drug." *Progress in Drug Research* 68:171–179. www.ncbi.nlm.nih.gov/pubmed/24941669.

Ley, Ruth E. 2016. "Gut Microbiota in 2015: Prevotella in the Gut: Choose Carefully." *Nature Reviews Gastroenterology & Hepatology* 13:69–70. http://dx.doi.org/10.1038 /nrgastro.2016.4.

Ley, Ruth E., et al. 2005. "Obesity Alters Gut Microbial Ecology." *Procedings of the National Academy of Sciences of the United States of America* 102 (31): 11070–11075. http://dx.doi.org/10.1073/pnas.0504978102.

Li, Endan, et al. 2013. "Ghrelin Directly Stimulates Adult Hippocampal Neurogenesis: Implications for Learning and Memory." *Endocrine Journal* 60 (6): 781–789. http:// dx.doi.org/10.1507/endocrj.EJ13-0008.

Li, Liaoliao, Zhi Wang, and Zhiyi Zuo. 2013. "Chronic Intermittent Fasting Improves Cognitive Functions and Brain Structures in Mice." *PLoS One* 8 (6): e66069. http:// dx.doi.org/10.1371/journal.pone.0066069.

Lichtenbelt, Wouter D. van Marken, et al. 2009. "Cold-Activated Brown Adipose Tissue in Healthy Men." *New England Journal of Medicine* 360:1500–1508. http://dx.doi .org/10.1056/NEJMoa0808718.

Lieber, C. S. 1987. "Microsomal Ethanol-Oxidizing System." *Europe PMC* 37 (1–2): 45–56. http://europepmc.org/abstract/med/3106031.

Lin, Hua V., et al. 2012. "Butyrate and Propionate Protect Against Diet-Induced Obesity and Regulate Gut Hormones via Free Fatty Acid Receptor 3-Independent Mechanisms." *PLoS One* 7 (4): e35240. http://dx.doi.org/10.1371/journal.pone .0035240.

Little, T. J., M. Horowitz, and C. Feinle-Bisset. 2005. "Role of Cholecystokinin in Appetite Control and Body Weight Regulation." *Obesity Reviews* 6 (4): 297–306. http://dx.doi .org/10.1111/j.1467-789X.2005.00212.x.

Liu, Simin, et al. 2003. "Relation Between Changes In Intakes of Dietary Fiber and Grain Products and Changes in Weight and Development of Obesity Among Middle-Aged Women." *American Journal of Clinical Nutrition* 78 (5): 920–927. http://ajcn .nutrition.org/content/78/5/920.long.

Lizcano, Fernando, and Guillermo Guzmán. 2014. "Estrogen Deficiency and the Origin of Obesity During Menopause." *BioMed Research International* 2014:757461. http:// dx.doi.org/10.1155/2014/757461.

Long, S. J., K. Hart, and L. M. Morgan. 2002. "The Ability of Habitual Exercise to Influence Appetite and Food Intake in Response to High- and Low-Energy Preloads in Man." *British Journal of Nutrition* 87 (5): 517–523. https://dx.doi.org/10.1079 /BJNBJN2002560.

Longhi, Silvia, and Giorgio Radetti. 2013. "Thyroid Function and Obesity." *Journal of Clinical Research in Pediatric Endocrinology* 5 (suppl. 1): 40–44. http://dx.doi .org/10.4274/Jcrpe.856.

Ludwig, David, Karen E. Peterson, and Steven L. Gortmaker. 2001. "Relation Between Consumption of Sugar-Sweetened Drinks and Childhood Obesity: A Prospective, Observational Analysis." *Lancet* 357 (9255): 505–508. www.researchgate.net/profile /Karen_Peterson3/publication/12101413_Relation_between_Consumption_of _Sugar_Sweetened_Drinks_and_Childhood_Obesity_A_Prospective_Observational _Study/links/0c96052b06a0e117c8000000.pdf.

Luppino, Floriana S., et al. 2010. "Overweight, Obesity, and Depression." *JAMA Psychiatry* 67 (3): 220–229. http://dx.doi.org/10.1001/archgenpsychiatry.2010.2.

Lutter, Michael, et al. 2008. "The Orexigenic Hormone Ghrelin Defends Against Depressive Symptoms of Chronic Stress." *Nature Neuroscience* 11:752–753. https://dx.doi.org/10.1038%2Fnn.2139.

Lynch, Christopher J., and Sean H. Adams. 2014. "Branched-Chain Amino Acids in Metabolic Signalling and Insulin Resistance." *Nature Reviews Endocrinology* 10:723–736. http://dx.doi.org/10.1038/nrendo.2014.171.

MacNeil, Douglas J. 2013. "The Role of Melanin-Concentrating Hormone and Its Receptors in Energy Homeostasis." *Frontiers in Endocrinology* 4:49. http://dx.doi.org/10.3389/fendo.2013.00049.

Maeda, Kazuhisa, and Yuji Matsuzawa. 2014. "Discovery of Adiponectin and Its Future Prospect." *Ommega*:1–9. www.ommegaonline.org/article-details/Discovery-of-Adiponectin-and-its-Future-Prospect/187.

Malaisse, W. J., et al. 1998. "Effects of Artificial Sweeteners on Insulin Release and Cationic Fluxes in Rat Pancreatic Islets." *Cellular Signalling* 10 (10): 727–733. www.ncbi.nlm.nih.gov/pubmed/9884024.

Maripuu, Martin, et al. 2016. "Relative Hypocortisolism Is Associated with Obesity and the Metabolic Syndrome in Recurrent Affective Disorders." *Journal of Affective Disorders* 204:187–196. http://dx.doi.org/10.1016/j.jad.2016.06.024.

Marzo, Vincenzo Di, and Isabel Matias. 2005. "Endocannabinoid Control of Food Intake and Energy Balance." *Nature Neuroscience* 8:585–589. http://dx.doi.org/10.1038/nn1457.

Massrieh, Wael. 2008. "Health Benefits of Omega-3 Fatty Acids from Neptune Krill Oil." *Lipid Technology* 20 (5): 108–111. http://dx.doi.org/10.1002/lite.200800022.

McCarty, Mark F., James J. DiNicolantonio, and James H. O'Keefe. 2015. "Capsaicin May Have Important Potential for Promoting Vascular and Metabolic Health." *Open Heart* 2 (1): e000262. http://dx.doi.org/10.1136/openhrt-2015-000262.

McMillen, I. Caroline, Clare L. Adam, and Beverly S. Mühlhäusler. 2005. "Early Origins of Obesity: Programming the Appetite Regulatory System." *Journal of Physiology* 565 (1): 9–17. http://dx.doi.org/10.1113/jphysiol.2004.081992.

Miller, Colette N., et al. 2012. "Estrogens, Inflammation and Obesity: An Overview." *Frontiers in Biology* 7 (1): 40–47. http://dx.doi.org/10.1007/s11515-011-1174-y.

Mistry, Anahita M., Andrew G. Swick, and Dale R. Romsos. 1997. "Leptin Rapidly Lowers Food Intake and Elevates Metabolic Rates in Lean and ob/ob Mice." *Journal of Nutrition* 127 (10): 2065–2072. http://jn.nutrition.org/content/127/10/2065.long.

Mohamed, Gamal A., et al. 2014. "Natural Anti-obesity Agents." *Bulletin of Faculty of Pharmacy, Cairo University* 52 (2): 269–284. http://dx.doi.org/10.1016/j.bfopcu.2014.05.001.

Moller, Niels, Peter O'Brien, and K. Sreekumaran Nair. 1998. "Disruption of the Relationship Between Fat Content and Leptin Levels with Aging in Humans." *Journal of Clinical Endocrinology and Metabolism*. http://dx.doi.org/10.1210/jcem.83.3.4620.

Monte, Suzanne M. de la, and Jack R. Wands. 2008. "Alzheimer's Disease Is Type 3 Diabetes–Evidence Reviewed." *Journal of Diabetes Science and Technology* 2 (6): 1101–1113. http://dx.doi.org/10.1177/193229680800200619.

Monte, Suzanne M. de la, et al. 2009. "Nitrosamine Exposure Exacerbates High Fat Diet-Mediated Type 2 Diabetes Mellitus, Non-alcoholic Steatohepatitis, and Neurodegeneration with Cognitive Impairment." *Molecular Neurodegeneration* 4:54. http://dx.doi.org/10.1186/1750-1326-4-54.

Morgan, Oliver W., et al. 2010. "Morbid Obesity as a Risk Factor for Hospitalization and Death Due to 2009 Pandemic Influenza A(H1N1) Disease." *PLoS One* 5 (3): e9694. http://dx.doi.org/10.1371/journal.pone.0009694.

Movahedi, Mohammad, et al. 2015. "Obesity, Aspirin, and Risk of Colorectal Cancer in Carriers of Hereditary Colorectal Cancer: A Prospective Investigation in the CAPP2 Study." *Journal of Clinical Oncology* 33 (21): 3591–3597. https://dx.doi.org/10.1200/JCO.2014.58.9952.

Mozaffarian, Dariush, et al. 2010. "*Trans*-Palmitoleic Acid, Metabolic Risk Factors, and New-Onset Diabetes in US Adults." *Annals of Internal Medicine* 153 (12): 790–799. http://dx.doi.org/10.7326/0003-4819-153-12-201012210-00005.

Mozaffarian, Dariush, et al. 2013. "*Trans*-Palmitoleic Acid, Other Dairy Fat Biomarkers, and Incident Diabetes." *American Journal of Clinical Nutrition* 97:854–861. https://cdn1.sph.harvard.edu/wp-content/uploads/sites/147/2012/10/2013_Am-J-Clin-Nutr_trans-Palmitoleic-acid_Mozaffarian.pdf.

Muoio, Deborah M., et al. 1997. "Leptin Directly Alters Lipid Partitioning in Skeletal Muscle." *Diabetes* 46 (8): 1360–1363. http://dx.doi.org/10.2337/diab.46.8.1360.

Naguib, Yousry M. A. 2000. "Antioxidant Activities of Astaxanthin and Related Carotenoids." *Journal of Agricultural and Food Chemistry* 48:1150–1154. http://dx.doi.org/10.1021/jf991106k.

Nakagawa, Yuko, et al. 2009. "Sweet Taste Receptor Expressed in Pancreatic β-Cells Activates the Calcium and Cyclic AMP Signaling Systems and Stimulates Insulin Secretion." *PLoS One* 4 (4): e5106. http://dx.doi.org/10.1371/journal.pone.0005106.

Negro, M., et al. 2008. "Branched-Chain Amino Acid Supplementation Does Not Enhance Athletic Performance but Affects Muscle Recovery and the Immune System." *Journal of Sports Medicine and Physical Fitness* 48 (3): 347–351. www.ncbi.nlm.nih.gov/pubmed/18974721.

Newgard, Christopher B., et al. 2009. "A Branched-Chain Amino Acid Related Metabolic Signature That Differentiates Obese and Lean Humans and Contributes to Insulin Resistance." *Cell Metabolism* 9:311–326. http://dx.doi.org/10.1016/j.cmet.2009.02.002.

Nguyen, Thi Loan Anh, et al. 2015. "How Informative Is the Mouse for Human Gut Microbiota Research?" *Disease Models and Mechanisms* 8:1–16. https://dx.doi.org/10.1242%2Fdmm.017400.

Ogden, Cynthia L., and Margaret D. Carroll. 2010. "Prevalence of Overweight, Obesity, and Extreme Obesity Among Adults: United States, Trends 1960–1962 Through 2007 2008." National Center for Health Statistics. www.cdc.gov/nchs/data/hestat/obesity_adult_07_08/obesity_adult_07_08.pdf.

Ollmann, Michael M., et al. 1997. "Antagonism of Central Melanocortin Receptors in Vitro and in Vivo by Agouti-Related Protein." *Science* 278:135–138. https://dx.doi.org/10.1126%2Fscience.278.5335.135.

Östman, E., et al. 2005. "Vinegar Supplementation Lowers Glucose and Insulin Responses and Increases Satiety After a Bread Meal in Healthy Subjects." *European Journal of Clinical Nutrition* 59:983–988. https://dx.doi.org/10.1038/sj.ejcn.1602197.

Page, Amanda J., et al. 2007. "Ghrelin Selectively Reduces Mechanosensitivity of Upper Gastrointestinal Vagal Afferents." *Gastrointestinal and Liver Physiology* 292 (5): G1376–G1384. https://dx.doi.org/10.1152%2Fajpgi.00536.2006.

Page, Kathleen A., et al. 2013. "Effects of Fructose vs. Glucose on Regional Cerebral Blood Flow in Brain Regions Involved with Appetite and Reward Pathways." *Journal of the American Medical Association* 309 (1):63–70. http://dx.doi.org/10.1001/jama.2012.116975.

Palmer, Biff F., and Deborah J. Clegg. 2015. "The Sexual Dimorphism of Obesity."
  Molecular and Cellular Endocrinology, 402:113–119. http://dx.doi.org/10.1016/j
  .mce.2014.11.029.
Pasarica, Magdalena, Scott Loiler, and Nikhil V. Dhurandhar. 2008. "Acute Effect of
  Infection by Adipogenic Human Adenovirus Ad36." *Archives of Virology* 153 (11):
  2097–2102. https://dx.doi.org/10.1007%2Fs00705-008-0219-2.
Payne, A. N., C. Chassard, and C. Lacroix. 2012. "Gut Microbial Adaptation to Dietary
  Consumption of Fructose, Artificial Sweeteners and Sugar Alcohols: Implications
  for Host–Microbe Interactions Contributing to Obesity." *Obesity Reviews* 13 (9):
  799–809. http://dx.doi.org/10.1111/j.1467-789X.2012.01009.x.
Pérez, Coralia, et al. 2004. "Leptin Impairs Insulin Signalling in Rat Adipocytes." *Diabetes*
  53 (10): 347–353. http://dx.doi.org/10.2337/diabetes.53.2.347.
Perry, B., and Y. Wang. 2012. "Appetite Regulation and Weight Control: The Role of Gut
  Hormones." *Nutrition & Diabetes* 2 (e26): 1–7. http://dx.doi.org/10.1038
  /nutd.2011.21.
Pocai, Alessandro. 2013. "Action and Therapeutic Potential of Oxyntomodulin." *Molecular
  Metabolism* 3 (3): 241–251. https://dx.doi.org/10.1016%2Fj.molmet.2013.12.001.
Poehlman, E. T., et al. 1986. "Genotype Dependency of Adaptation in Adipose Tissue
  Metabolism After Short-Term Overfeeding." *American Journal of Physiology—
  Endocrinology and Metabolism* 250 (4): E480–E485. http://ajpendo.physiology.org
  /content/250/4/E480.
Ponterio, Eleonora, and Lucio Gnessi. 2015. "Adenovirus 36 and Obesity: An Overview."
  Viruses 7 (7): 3719–3740. https://dx.doi.org/10.3390%2Fv7072787.
Potenza, Matthew, Michael Via, and Robert Yanagisawa. 2009. "Excess Thyroid Hormone
  and Carbohydrate Metabolism." *Endocrine Practice* 15 (2): 254–262. http://dx.doi
  .org/10.4158/EP.15.3.254.
Prados, Andreu. 2016. "The Latest Advances Regarding the Link Between Prevotella
  Genus, Diet and Its Impact on Host Health." *Gut Microbiota Research and Practice*,
  February 11. www.gutmicrobiotaforhealth.com/en/the-latest-advances-regarding-the
  -link-between-prevotella-genus-diet-and-its-impact-on-host-health/.
Pratt, Laura A., and Debra J. Brody. 2014. "Depression and Obesity in the US Adult
  Household Population, 2005–2010." National Center for Health Statistics (CDC):
  1–8. www.cdc.gov/nchs/data/databriefs/db167.pdf.
Prinz, Patricia N., et al. 2009. "Effect of Alcohol on Sleep and Nighttime Plasma Growth
  Hormone and Cortisol Concentrations." *Journal of Clinical Endocrinology and
  Metabolism* 51 (4). http://dx.doi.org/10.1210/jcem-51-4-759.
Pritchard, L. E., A. V. Turnbull, and A. White. 2002. "Pro-opiomelanocortin Processing
  in the Hypothalamus: Impact on Melanocortin Signalling and Obesity." *Journal of
  Endocrinology* 172:411–421. http://joe.endocrinology-journals.org/content/172/3
  /411.full.pdf.
Qian, Weiyun, et al. 2014. "Decreased Circulating Levels of Oxytocin in Obesity and
  Newly Diagnosed Type 2 Diabetic Patients." *Journal of Clinical Endocrinology and
  Metabolism* 99 (12): 4683–4689. http://dx.doi.org/10.1210/jc.2014-2206.
Raimondi, Stefano, et al. 2016. "Conjugated Linoleic Acid Production by Bifidobacteria:
  Screening, Kinetic, and Composition." *BioMed Research International* 2016:8654317.
  http://dx.doi.org/10.1155/2016/8654317.
Ramachandrappa, Shwetha, and I. Sadaf Farooqi. 2011. "Genetic Approaches to
  Understanding Human Obesity." *Journal of Clinical Investigation* 121 (6): 2080–2086.
  https://doi.org/10.1172/JCI46044.

Rasmussen, Michael Højby. 2010. "Obesity, Growth Hormone and Weight Loss." *Molecular and Cellular Endocrinology* 316 (2): 147–153. http://dx.doi.org/10.1016/j .mce.2009.08.017.

Reinehr, T., et al. 2006. "Pancreatic Polypeptide in Obese Children Before and After Weight Loss." *International Journal of Obesity* 30:1476–1481. http://dx.doi.org/10 .1038/sj.ijo.0803393.

Richard, D., Q. Huang, and E. Timofeeva. 2000. "The Corticotropin-Releasing Hormone System in the Regulation of Energy Balance in Obesity." *International Journal of Obesity* 24 (2): S36–S39. www.nature.com/ijo/journal/v24/n2s/pdf/0801275a .pdf?origin=publication_detail.

Ridaura, Vanessa K., et al. 2013. "Gut Microbiota from Twins Discordant for Obesity Modulate Metabolism in Mice." *Science* 342:1079–1089. http://dx.doi.org/10.1126 /science.1241214.

Rosenbaum, M., et al. 2011. "Effects of Gender, Body Composition, and Menopause on Plasma Concentrations of Leptin." *Journal of Clinical Endocrinology and Metabolism* 81 (9): 3424–3427. http://dx.doi.org/10.1210/jcem.81.9.8784109.

Sainsbury, A., et al. 1996. "Intracerebroventricular Administration of Neuropeptide Y to Normal Rats Increases Obese Gene Expression in White Adipose Tissue. *Diabetologia* 39:353–356. https://dx.doi.org/10.1007/BF00418353.

Saito, Masayuki, and Takeshi Yoneshiro. 2013. "Capsinoids and Related Food Ingredients Activating Brown Fat Thermogenesis and Reducing Body Fat in Humans." *Current Opinion in Lipidology* 24 (1): 71–77. https://dx.doi.org/10.1097/MOL .0b013e32835a4f40.

Sanz, Yolanda, et al. 2015. "Understanding the Role of Gut Microbiome in Metabolic Disease Risk." *Pediatric Research* 77:236–244. http://dx.doi.org/10.1038/pr.2014 .170.

Saygın, M., et al. 2016. "The Impact of High Fructose on Cardiovascular System: Role of α-Lipoic Acid." *Human and Experimental Toxicology* 35 (2): 194–204. http://dx.doi .org/10.1177/0960327115579431.

Sears, Barry, and Camillo Ricordi. 2010. "Anti-inflammatory Nutrition as a Pharmacological Approach to Treat Obesity." *Journal of Obesity* 2011:431985. https:// dx.doi.org/10.1155%2F2011%2F431985.

Seino, Yutaka, and Daisuke Yabe. 2013. "Glucose-Dependent Insulinotropic Polypeptide and Glucagon-Like Peptide-1: Incretin Actions Beyond the Pancreas." *Journal of Diabetes Investigation* 4 (2): 108–130. http://dx.doi.org/10.1111/jdi .12065.

Sender, Ron, Shai Fuchs, and Ron Milo. 2016. "Revised Estimates for the Number of Human and Bacteria Cells in the Body." *PLOS Biology* 14 (8): e1002533. http:// dx.doi.org/10.1371/journal.pbio.1002533.

Sergent, Thérèse, et al. 2012. "Phenolic Compounds and Plant Extracts as Potential Natural Anti-obesity Substances." *Food Chemistry* 135 (1): 68–73. http://dx.doi.org/10.1016/j .foodchem.2012.04.074.

She, Pengxiang, et al. 2007. "Obesity-Related Elevations in Plasma Leucine Are Associated with Alterations in Enzymes Involved in Branched Chain Amino Acid (BCAA) Metabolism." *American Journal of Physiology—Endocrinology and Metabolism* 293 (6): E1552–E1563. https://dx.doi.org/10.1152%2Fajpendo.00134.2007.

Shehzad, Adeeb, et al. 2011. "New Mechanisms and the Anti-inflammatory Role of Curcumin in Obesity and Obesity-Related Metabolic Diseases." *European Journal of Nutrition* 50 (3): 151–161. http://dx.doi.org/10.1007%2Fs00394-011-0188-1.

Shimabukuro, Michio, et al. 1997. "Direct Antidiabetic Effect of Leptin Through Triglyceride Depletion." *Proceedings of the National Academy of Sciences* 94:4637–4641. www.researchgate.net/profile/Guoxun_Chen/publication/45072886_Direct _Antidiabetic_Effect_of_Leptin_through_Triglyceride_Depletion_of_Tissues /links/0f3175320e32091374000000.pdf.

Shutter, J. R., et al. 1997. "Hypothalamic Expression of ART, a Novel Gene Related to Agouti, Is Up-Regulated in Obese and Diabetic Mutant Mice." *Genes & Development* 11:593–602. https://dx.doi.org/10.1101%2Fgad.11.5.593.

Singh, Madhuri, and Kasturi Mukhopadhyay. 2014. "Alpha-Melanocyte Stimulating Hormone: An Emerging Anti-inflammatory Antimicrobial Peptide." *BioMed Research International* 2014:874610. http://dx.doi.org/10.1155/2014/874610.

Skibicka, Karolina P., et al. 2012. "Role of Ghrelin in Food Reward: Impact of Ghrelin on Sucrose Self-administration and Mesolimbic Dopamine and Acetylcholine Receptor Gene Expression." *Addiction Biology* 17 (1): 95–107. https://dx.doi.org/10.1111%2Fj .1369-1600.2010.00294.x.

Slavin, J. L. 2005. "Dietary Fiber and Body Weight." *Nutrition* 21 (3): 411–418. www .ncbi.nlm.nih.gov/pubmed/15797686.

Smith, Gordon I., et al. 2016. "High-Protein Intake During Weight Loss Therapy Eliminates the Weight-Loss-Induced Improvement in Insulin Action in Obese Postmenopausal Women." *Cell Reports* 17 (3): 849–861. http://dx.doi.org/10.1016/j .celrep.2016.09.047.

Spector, Tim. 2015. *The Diet Myth*. London: Weidenfeld & Nicolson.

Spiegel, Karine, et al. 2005. "Sleep Loss: A Novel Risk Factor for Insulin Resistance and Type 2 Diabetes." *Journal of Applied Physiology* 99 (5): 2008–2019. http://dx.doi .org/10.1152/japplphysiol.00660.2005.

Spiegel, Karine, Rachel Leproult, and Eve Van Cauter. 1999. "Impact of Sleep Debt on Metabolic and Endocrine Function." *Lancet* 354 (9188): 1435–1439. http://dx.doi .org/10.1016/S0140-6736(99)01376-8.

Stanhope, Kimber L., et al. 2009. "Consuming Fructose-Sweetened, Not Glucose-Sweetened, Beverages Increases Visceral Adiposity and Lipids and Decreases Insulin Sensitivity in Overweight/Obese Humans." *Journal of Clinical Investigation* 119 (5): 1322–1334. http://dx.doi.org/10.1172/JCI37385.

Steinert, Robert E., et al. 2011. "Effects of Carbohydrate Sugars and Artificial Sweeteners on Appetite and the Secretion of Gastrointestinal Satiety Peptides." *British Journal of Nutrition* 105 (9): 1320–1328. http://dx.doi.org/10.1017/S000711451000512X.

Stewart, W. K., and Laura W. Fleming. 1973. "Features of a Successful Therapeutic Fast of 382 Days' Duration." *Postgraduate Medical Journal* 49:203–209. www.ncbi.nlm.nih .gov/pmc/articles/PMC2495396/pdf/postmedj00315-0056.pdf.

Stunkard, Albert J., Myles S. Faith, and Kelly C. Allison. 2003. "Depression and Obesity." *Biological Psychiatry* 54 (3): 330–337. http://dx.doi.org/10.1016/0006-3223(03) 00608-5.

Suez, Jotham, et al. 2014. "Artificial Sweeteners Induce Glucose Intolerance by Altering the Gut Microbiota." *Nature* 514:181–186. http://dx.doi.org/10.1038/nature13793.

SugarStacker. 2016. *Sugar Stacks*. www.sugarstacks.com/fruits.htm.

Surwit, Richard S., et al. 2010. "Plasma Epinephrine Predicts Fasting Glucose in Centrally Obese African-American Women." *Obesity* 18:1683–1687. http://dx.doi.org/10.1038 /oby.2010.43.

Susuki, Michitaka, et al. 2011. "Lipid Droplets: Size Matters." *Journal of Electron Microscopy* 60 (suppl. 1): S101–S116. http://dx.doi.org/10.1093/jmicro/dfr016.

Suter, Paolo M., and Angelo Tremblay. 2008. "Is Alcohol Consumption a Risk Factor for Weight Gain and Obesity?" *Critical Reviews in Clinical Laboratory Sciences* 42 (3): 197–227. http://dx.doi.org/10.1080/10408360590913542.

Sutin, Angelina R., et al. 2011. "Personality and Obesity Across the Adult Lifespan." *Journal of Personality and Social Psychology* 101 (3): 579–592. https://dx.doi.org/10.1037%2Fa0024286.

Swithers, Susan E. 2013. "Artificial Sweeteners Produce the Counterintuitive Effect of Inducing Metabolic Derangements." *Trends in Endocrinology & Metabolism* 24 (9): 431–441. http://dx.doi.org/10.1016/j.tem.2013.05.005.

Taheri, Shahrad, et al. 2004. "Short Sleep Duration Is Associated with Reduced Leptin, Elevated Ghrelin, and Increased Body Mass Index." *PLOS Medicine* 1 (3): 210–217. https://dx.doi.org/10.1371%2Fjournal.pmed.0010062.

Tamega, Andréia de Almeida, et al. 2010. "Association Between Skin Tags and Insulin Resistance." *Anais Brasileiros de Dermatologia* 85 (1): 25–31. http://dx.doi. org/10.1590/S0365-05962010000100003.

Tang, Hong, et al. 2016. "Irisin Inhibits Hepatic Cholesterol Synthesis via AMPK-SREBP2 Signaling." *EBioMedicine* 6:139–148. http://dx.doi .org/10.1016/j.ebiom.2016.02.041.

Teff, Karen L., et al. 2004. "Dietary Fructose Reduces Circulating Insulin and Leptin, Attenuates Postprandial Suppression of Ghrelin, and Increases Triglycerides in Women." *Journal of Clinical Endocrinology and Metabolism* 89 (6): 2963–2972. http:// dx.doi.org/10.1210/jc.2003-031855#sthash.hNiPziUg.i3rggj8m.dpuf.

Tong, Ming, et al. 2009. "Nitrosamine Exposure Causes Insulin Resistance Diseases: Relevance to Type 2 Diabetes Mellitus, Non-alcoholic Steatohepatitis, and Alzheimer's Disease." *Journal of Alzheimer's Disease* 17 (4): 827–844. www.ncbi.nlm.nih.gov/pmc /articles/PMC2952429/.

Tong, Ming, Lisa Longato, and Suzanne M. de la Monte. 2010. "Early Limited Nitrosamine Exposures Exacerbate High Fat Diet-Mediated Type 2 Diabetes and Neurodegeneration." *BMC Endocrine Disorders* 10 (4): 1–16. http://dx.doi.org/10 .1186/1472-6823-10-4.

Trümper, Andrea, et al. 2009. "Glucose-Dependent Insulinotropic Polypeptide Is a Growth Factor for β (INS-1) Cells by Pleiotropic Signaling." *Molecular Endocrinology* 15 (9): 1559–1570. http://dx.doi.org/10.1210/mend.15.9.0688.

Turnbaugh, Peter J., et al. 2006. "An Obesity-Associated Gut Microbiome with Increased Capacity for Energy Harvest." *Nature* 444:1027–1031. http://dx.doi.org/10.1038 /nature05414.

Vaisse, Christian, et al. 1998. "A Frameshift Mutation in Human MC4R Is Associated with a Dominant Form of Obesity." *Nature Genetics* 20 (2): 113–114. https://dx.doi .org/10.1038/2407.

VanHelder, T., J. D. Symons, and M. W. Radomski. 1993. "Effects of Sleep Deprivation and Exercise on Glucose Tolerance." *Aviation, Space, and Environmental Medicine* 64 (6): 487–492. www.ncbi.nlm.nih.gov/pubmed/8338493.

Vehapoğlu, Aysel, Serdar Türkmen, and Sule Terzioglu. 2015. "Alpha-Melanocyte-Stimulating Hormone and Agouti-Related Protein: Do They Play a Role in Appetite Regulation in Childhood Obesity?" *Journal of Clinical Research in Pediatric Endocrinology* 8 (1): 40–47. http://dx.doi.org/10.4274/jcrpe.2136.

Vgontzas, A. N., E. O. Bixler, and G. P. Chrousos. 2003. "Metabolic Disturbances in Obesity Versus Sleep Apnoea: The Importance of Visceral Obesity and Insulin Resistance." *Journal of Internal Medicine* 254 (1): 32–44. http://dx.doi.org/10 .1046/j.1365-2796.2003.01177.x.

Vicennati, Valentina, et al. 2002. "Response of the Hypothalamic-Pituitary-Adrenocortical Axis to High-Protein/Fat and High-Carbohydrate Meals in Women with Different Obesity Phenotypes." *Journal of Clinical Endocrinology & Metabolism* 97 (8): 3984–3988. http://dx.doi.org/10.1210/jcem.87.8.8718.

Vickers, Mark H., et al. 2000. "Fetal Origins of Hyperphagia, Obesity, and Hypertension and Postnatal Amplification by Hypercaloric Nutrition." *American Journal of Physiology and Endocrinology* 279:E83–E87. www.researchgate.net/profile/Bernhard_Breier/publication/12426537_Fetal_origins_of_hyperphagia_obesity_and_hypertension_and_postnatal_amplification_by_hypercaloric_nutrition_Am_J_Physiol_279E83-E87/links/574ca21d08ae8bc5d15a4372.pdf.

Villareal, Dennis T., et al. 2005. "Obesity in Older Adults: Technical Review and Position Statement of the American Society for Nutrition and NAASO, the Obesity Society." *American Journal of Clinical Nutrition* 82 (5): 923–934. http://ajcn.nutrition.org/content/82/5/923.full.

Vliet-Ostaptchouk, Jana V. van, et al. 2008. "Polymorphisms of the TUB Gene Are Associated with Body Composition and Eating Behavior in Middle-Aged Women." *PLoS One* 3 (1): 1–8. http://dx.doi.org/10.1371/journal.pone.0001405.

Vrieze, Anne, et al. 2012. "Transfer of Intestinal Microbiota from Lean Donors Increases Insulin Sensitivity in Individuals with Metabolic Syndrome." *Gastroenterology* 143 (4): 913–916. http://dx.doi.org/10.1053/j.gastro.2012.06.031.

Vrieze, Anne, et al. 2013. "Chapter 3: The Therapeutic Potential of Manipulating Gut Microbiota in Obesity and Type 2 Diabetes Mellitus." http://dare.uva.nl/document/2/119966.

Walleghen, E. L. Van, et al. 2007. "Habitual Physical Activity Differentially Affects Acute and Short-Term Energy Intake Regulation in Young and Older Adults." *International Journal of Obesity* 31:1277–1285. http://dx.doi.org/10.1038/sj.ijo.0803579.

Wallerius, S., et al. 2003. "Rise in Morning Saliva Cortisol Is Associated with Abdominal Obesity in Men: A Preliminary Report." *Journal of Endocrinological Investigation* 26 (7): 616–619. https://dx.doi.org/10.1007/BF03347017.

Wang, S., et al. 2015. "Resveratrol Induces Brown-Like Adipocyte Formation in White Fat Through Activation of AMP-Activated Protein Kinase (AMPK) α1." *International Journal of Obesity* 39:967–976. http://dx.doi.org/10.1038/ijo.2015.23.

Wang, Trevor L., Tanya Ya. Bogracheva, and Cliff L. Hedley. 1998. "Starch: as Simple as A, B, C?" *Journal of Experimental Botany* 49 (320): 481–502. http://jxb.oxfordjournals.org/content/49/320/481.full.pdf.

Weigle, David S., et al. 2005. "A High-Protein Diet Induces Sustained Reductions in Appetite, Ad Libitum Caloric Intake, and Body Weight Despite Compensatory Changes in Diurnal Plasma Leptin and Ghrelin Concentrations." *American Society for Clinical Nutrition* 82 (1): 41–48. http://ajcn.nutrition.org/content/82/1/41.full.pdf+html.

Westerterp-Plantenga, Margriet S., and Christianne R. T. Verwegen. 1999. "The Appetizing Effect of an Apéritif in Overweight and Normal-Weight Humans." *American Journal of Clinical Nutrition* 69 (2): 205–212. http://ajcn.nutrition.org/content/69/2/205.full.pdf.

Westerterp-Plantenga, Margriet S., et al. 2004. "High Protein Intake Sustains Weight Maintenance After Body Weight Loss in Humans." *International Journal of Obesity* 28:57–64. http://dx.doi.org/10.1038/sj.ijo.0802461.

Willie, Jon T., et al. 2001. "To Eat or Sleep? Orexin in the Regulation of Feeding and Wakefulness." *Neuroscience* 24:429–458. http://dx.doi.org/10.1146/annurev .neuro.24.1.429.

Wise, Paul M., et al. 2015. "Reduced Dietary Intake of Simple Sugars Alters Perceived Sweet Taste Intensity but Not Perceived Pleasantness." *American Society for Nutrition* 103 (1): 50–60. http://dx.doi.org/10.3945/ajcn.115.112300.

Wolfram, Swen, Ying Wang, and Frank Thielecke. 2006. "Anti-obesity Effects of Green Tea: From Bedside to Bench." *Molecular Nutrition & Food Research* 50 (2): 176–187. https://dx.doi.org/10.1002/mnfr.200500102.

Wrann, Christiane D. 2015. "FNDC5/Irisin—Their Role in the Nervous System and as a Mediator for Beneficial Effects of Exercise on the Brain." *Brain Plasticity* 1 (1): 55–61. http://dx.doi.org/10.3233/BPL-150019.

Wu, Gary D., et al. 2011. "Linking Long-Term Dietary Patterns with Gut Microbial Enterotypes." *Science* 334:105–108. https://dx.doi.org/10.1126/science.1208344.

Xia, Charley, et al. 2016. "Pedigree- and SNP-Associated Genetics and Recent Environment Are the Major Contributors to Anthropometric and Cardiometabolic Trait Variation." *PLOS Genetics* 13 (2): e1006608. http://dx.doi.org/10.1371/journal.pgen.1005804.

Xu, Xiangbin, et al. 2008. "Molecular Mechanisms of Ghrelin-Mediated Endothelial Nitric Oxide Synthase Activation." *Endocrinology* 149 (8): 4183–4192. http://dx.doi.org /10.1210/en.2008-0255.

Xu, Zhenjiang, and Rob Knight. 2015. "Dietary Effects on Human Gut Microbiome Diversity." *British Journal of Nutrition* 113 (suppl. 1): S1–S5. https://doi.org/10.1017 /S0007114514004127.

Yadav, Amita, et al. 2013. "Role of Leptin and Adiponectin in Insulin Resistance." *Clinica Chimica Acta* 417:80–84. http://dx.doi.org/10.1016/j.cca.2012.12.007.

Yamagishi, Kazumasa, et al. 2010. "Dietary Intake of Saturated Fatty Acids and Mortality from Cardiovascular Disease in Japanese: The Japan Collaborative Cohort Study for Evaluation of Cancer Risk (JACC) Study." *American Journal of Clinical Nutrition* 92 (4): 759–765. http://dx.doi.org/10.3945/ajcn.2009.29146.

Yamashita, K., et al. 2012. "Association of Coffee Consumption with Serum Adiponectin, Leptin, Inflammation and Metabolic Markers in Japanese Workers: A Cross-Sectional Study." *Nutrition and Diabetes* 2: e33. http://dx.doi.org/10.1038/nutd.2012.6.

Yin, Jun, Huili Xing, and Jianping Ye. 2008. "Efficacy of Berberine in Patients with Type 2 Diabetes." *Metabolism* 57 (5): 712–717. https://dx.doi.org/10.1016%2Fj.metabol .2008.01.013.

Yin, Yue, Yin Li, and Weizhen Zhang. 2014. "The Growth Hormone Secretagogue Receptor: Its Intracellular Signaling and Regulation." *International Journal of Molecular Sciences* 15 (3): 4837–4855. https://dx.doi.org/10.3390%2Fijms15034837.

Yoo, Sae-Rom, et al. 2013. "Probiotics L. plantarum and L. curvatus in Combination Alter Hepatic Lipid Metabolism and Suppress Diet-Induced Obesity." *Obesity* 21 (12): 2571–2578. http://dx.doi.org/10.1002/oby.20428.

Yoshinari, Orie, Yoshiaki Shiojima, and Kiharu Igarashi. 2012. "Anti-obesity Effects of Onion Extract in Zucker Diabetic Fatty Rats." *Nutrients* 4 (10): 1518–1526. http:// dx.doi.org/10.3390/nu4101518.

You, Tongjian, et al. 2013. "Effects of Exercise Training on Chronic Inflammation in Obesity: Current Evidence and Potential Mechanisms." *Sports Medicine* 43 (4): 243–256. https://dx.doi.org/10.1007/s40279-013-0023-3.

Zhang, Hai, et al. 2013. "Treatment of Obesity and Diabetes Using Oxytocin or Analogs in Patients and Mouse Models." *PLoS One* 8 (5): e61477. https://dx.doi .org/10.1371%2Fjournal.pone.0061477.

Zhang, Yuan, et al. 2016. "Irisin Exerts Dual Effects on Browning and Adipogenesis of Human White Adipocytes." *American Journal of Physiology—Endocrinology and Metabolism* 311 (2): E530–E541. http://dx.doi.org/10.1152/ajpendo.00094.2016.

Zhao, Lu, et al. 2015. "Muscadine Grape Seed Oil as a Novel Source of Tocotrienols to Reduce Adipogenesis and Adipocyte Inflammation." *Food and Function* 6:2293–2302. http://dx.doi.org/10.1039/C5FO00261C.

Zhao, Xingrong, et al. 2012. "The Obesity and Fatty Liver Are Reduced by Plant-Derived Pediococcus pentosaceus LP28 in High Fat Diet-Induced Obese Mice." *PLoS One* 7 (2): e30696. http://dx.doi.org/10.1371/journal.pone.0030696.

Ziegler, Michael G., et al. 2011. "Endogenous Epinephrine Protects Against Obesity Induced Insulin Resistance." *Autonomic Neuroscience: Basic and Clinical* 162 (1–2): 32–34. https://dx.doi.org/10.1016%2Fj.autneu.2011.01.009.

Zocco, M. A., et al. 2007. "*Bacteroides thetaiotaomicron* in the Gut: Molecular Aspects of Their Interaction." *Digestive and Liver Disease* 39:707–712. http://dx.doi.org /10.1016/j.dld.2007.04.003.

INDEX